T0210813

Advancing Environmental Justice for Marginalized Communities in India

This interdisciplinary collection examines social equity and environmental justice in India. It assesses the effectiveness of environmental policies and institutions in rendering justice for marginalized communities while ensuring protection of the environment. It also analyses the influence of the neoliberal state and its political economies on the development and outcomes of these policies and institutions.

The book provides a unique perspective on environmental justice because of its consistent emphasis on social justice, rather than the prevailing predominant analyses from legal or environmental perspectives. It explores the themes of effectiveness and equity as they pertain to public policy instruments, such as environmental impact assessment, environmental licensing and enforcement, public hearings and environmental activism strategies. The four interlinked dimensions of environmental justice, namely recognitional justice, procedural justice, distributive justice and restorative justice, provide the core of the book's conceptual framework. The contributions draw on ideas and methods from development studies, environmental geography, environmental law and policy, natural resource management, public administration and political economy. The book concludes by considering planning, policy and institutional reforms and community-based initiatives that are needed to promote and protect environmental justice in India.

Offering an important reference for researchers and scholars, this book will appeal to those in law, geography, environmental studies, natural resource management, development studies, sociology and political science. It will also be of interest to community-based researchers, environmentalists and other civil society activists, natural resource managers and policy makers.

Alan P. Diduck is a Professor and the Department Chair of Environmental Studies and Sciences at the University of Winnipeg. Before joining the university, he was a lawyer and executive director of a social profit organization providing public legal education and information services. His research deals with community engagement in environmental governance, the learning implications of involvement and the consequences for social aspects of sustainability, such as adaptive capacity and environmental justice.

Kirit Patel is an Associate Professor and Program Chair of International Development Studies at Menno Simons College, affiliated with the University of Winnipeg and Canadian Mennonite University. As an academic, policy analyst and development professional, Dr Patel's teaching and research focus on environmental and social justice, sustainable food systems and nutrition security, agrarian change and rural–urban migration, agrobiodiversity conservation, Indigenous knowledge systems and the governance of common property resources in South Asia.

Aruna Kumar Malik was an Assistant Professor of Political Science at Gujarat National Law University. He was a recipient of the Queen Elizabeth II Diamond Jubilee International Scholarship and was a visiting scholar at the University of Winnipeg, Canada, in 2018–2019.

Advancing Environmental Justice for Marginalized Communities in India

Progress, Challenges and Opportunities

Edited by Alan P. Diduck, Kirit Patel and Aruna Kumar Malik

Routledge
Taylor & Francis Group

LONDON AND NEW YORK

First published 2022
by Routledge
2 Park Square, Milton Park, Abingdon, Oxon OX14 4RN

and by Routledge
605 Third Avenue, New York, NY 10158

Routledge is an imprint of the Taylor & Francis Group, an informa business

British Library Cataloguing-in-Publication Data
A catalogue record for this book is available from the British Library

Library of Congress Cataloging-in-Publication Data
Names: Diduck, Alan, editor. | Patel, Kirit, editor. | Malik, Aruna Kumar, editor.
Title: Advancing environmental justice for marginalized communities in India : progress, challenges and opportunities / edited by Alan P. Diduck, Kirit Patel, Aruna Kumar Malik.
Description: Abingdon, Oxon ; New York, NY : Routledge, 2022. | Includes bibliographical references and index.
Identifiers: LCCN 2021013899 (print) | LCCN 2021013900 (ebook) | ISBN 9780367692810 (hardback) | ISBN 9780367692827 (paperback) | ISBN 9781003141228 (ebook)
Subjects: LCSH: Environmental justice--India. | Environmental law--India.
Classification: LCC GE240.I4 A38 2022 (print) | LCC GE240.I4 (ebook) | DDC 363.7/052--dc23
LC record available at https://lccn.loc.gov/2021013899
LC ebook record available at https://lccn.loc.gov/2021013900

ISBN: 978-0-367-69281-0 (hbk)
ISBN: 978-0-367-69282-7 (pbk)
ISBN: 978-1-003-14122-8 (ebk)

DOI: 10.4324/9781003141228

Typeset in Bembo
by KnowledgeWorks Global Ltd.

Dedicated to the memory of our friend and colleague, Aruna Kumar Malik (1981–2021), who played a pivotal role in the book, from conception to final submission and publication.

Alan P. Diduck
Kirit Patel

Contents

PART I
Introduction

PART II
Economic, policy and institutional context

Figures

Tables

Contributors

J. Mark Baker is a Professor in the Department of Politics at Humboldt State University, California, where he also leads the Environment and Community MA in Social Science graduate programme. He is the author of *The Kuhls of Kangra: Community-Managed Irrigation in the Western Himalaya* (University of Washington Press, 2005) and the co-author of *Community Forestry in the United States: Learning from the Past, Crafting the Future* (Island Press, 2003).

Carinne Bétournay is a graduate of the Human Rights programme at the University of Winnipeg. Carinne has extensive professional, academic and volunteer experience in the international field: an internship with Makerere University in Uganda, mentorship in the Dominican Republic leading development projects followed by volunteer work in Haiti. While in university, Carinne travelled to Botswana and India to research the social determinants of health and restorative justice. Carinne anticipates working as an advocate for victims of gender-based violence and advancing the principles of environmental justice.

Alan P. Diduck is a Professor and the Department Chair of Environmental Studies and Sciences at the University of Winnipeg. Before joining the university, he was a lawyer and executive director of a social profit organization providing public legal education and information services. His research deals with community engagement in environmental governance, the learning implications of involvement and the consequences for social aspects of sustainability, such as adaptive capacity and environmental justice.

Ariane Dilay is a graduate of the Master of Natural Resources Management programme at the Natural Resources Institute, University of Manitoba. She has a Bachelor of Science with Honours in Environmental Studies and Sciences from the University of Winnipeg. Ariane is a recipient of a Social Sciences and Humanities Research Council award and has conducted research on environmental justice in India for both her honours' and master's theses.

Esther Edwards, now retired, remains an affiliated member of the Hazard, Risk and Disaster Research Group at Bath Spa University. She is a technical specialist in the use of satellite remote sensing, geographical information systems and environmental management. She has extensive expertise in airborne remote sensing for clients in Africa, Europe and South America. Her research focused on the use of mapping tools to understand societal behaviours in the context of environmental hazards. For years, she co-led collaborative research and teaching activities to the Indian Himalayas.

Bryce Gallant holds an MA in Gender and Development from the Institute of Development Studies at the University of Sussex. Her work has focused on rural development, environmental justice, urban agriculture and gender and inclusion. Currently she is a Gender, Youth and Inclusion Consultant with the International Water Management Institute in Columbo, Sri Lanka.

James Gardner is an Emeritus Professor at the University of Manitoba and Adjunct Professor of Geography at the University of Victoria. He lives in Victoria, British Columbia. His areas of expertise are in earth system sciences and their implications for hazard and disaster management, with a focus on mountain regions. He has 35 years' experience working on projects across the Himalaya, from Pakistan through India and Nepal to southwest China.

Debayan Gupta is a lawyer, graduating from the National Law University of Odisha in 2017. He has worked with the Centre for Policy Research–Namati Environmental Justice Program in New Delhi since 2017. He has a keen interest in environmental law and human rights, and in researching ways to make the law simpler and clearer.

Mahabaleshwar Hegde is the Head Researcher at the Centre for Policy Research–Namati Environmental Justice Program in New Delhi, India. In 2019–2020, he held a fellowship from the Queen Elizabeth II Diamond Jubilee Advanced Scholars Program, which supported a postdoctoral position at the University of Winnipeg in which he studied the role and impact of environmental litigation in India.

Shubham Janghu is a Master of Laws (LLM) graduate (2019–2020) from the University of Cambridge, where he specialized in international law. He completed his undergraduate law degree (BBA LLB) from O.P. Jindal Global University with a bronze medal. He is passionate about the law and has authored research publications on issues such as environmental law and human rights law.

Richard Johnson is the Director of Bath Spa University's Hazard, Risk and Disaster Research Group. He is a geomorphologist with commercial and academic expertise. His research focuses on upland/mountain sediment systems in the English Lake District, multi-disciplinary assessments of flood impacts/chronologies and disaster risk reduction in the Indian

Himalayas. Contributions include the film *Pathways to Resilience* (https://vimeo.com/285841577) and a science policy brief on disaster risk reduction in the Himalayan region (https://doi.org/10.17870/bathspa.c.5036990.v1).

Arpitha Kodiveri is a Hans Kelsen Fellow and doctoral researcher in law at the European University Institute. Her research examines land conflicts and legal mobilization by forest-dwelling communities in the mining areas of Odisha, India. She has previously worked as an environmental lawyer supporting forest-dwelling communities in their struggles to assert their forest rights. She has an LLM from the University of California Berkeley as a Fulbright Nehru Fellow and a BSL LLB from ILS Law College, Pune.

Avery Letkemann is a Research Coordinator and Senior Research Assistant for the Environment and Society Research Group at the University of Winnipeg. A graduate of the Department of Environmental Studies and Sciences, Letkemann's work focuses on environmental justice and resource governance. She has held positions as the Environmental Ethics Director for the University of Winnipeg's Student's Association and an advisory position with the City of Winnipeg as it developed a climate change action plan.

Aruna Kumar Malik was an Assistant Professor of Political Science at Gujarat National Law University. He was a recipient of the Queen Elizabeth II Diamond Jubilee International Scholarship and was a visiting scholar at the University of Winnipeg, Canada, in 2018–2019.

Avinash Mathews is an Associate at Shardul Amarchand Mangaldas, a leading Indian law firm. He is a gold medallist from the Jindal Global Law School, having completed a bachelor's degree in economics (Hons.) at St. Stephen's College, Delhi. He is also the founder of CEDE, a network of lawyers who provide formalized internship, mentorship and employment opportunities for socially marginalized groups.

Gayathri D. Naik is a Ph.D. candidate and a Commonwealth Scholar at the School of Law, SOAS (School of Oriental and African Studies), University of London. She holds a BAL/LLB (Topper, Mahatma Gandhi University, Kerala), an LLM (Gold Medallist, South Asian University, New Delhi) and an MPA (Indira Gandhi National Open University, New Delhi). She is a Junior Fellow in Climate Change Law and Policy at the Global Research Network and a member of the Law, Environment and Development Centre, SOAS, and several IUCN specialist groups.

Himanshu Pabreja graduated from Gujarat National Law University, India in 2019. He is currently working for a legal aid clinic engaged in grassroots environmental litigation and policy advisory. He has an avid interest in environmental law and policy making, among other areas of law that he is practicing in as a lawyer. He aims to bring about equitable environmental governance through the force of law.

Sangram Sinh Parab is an alumnus of Jindal Global Law School. He has worked as a law clerk and research assistant to the incumbent Chief Justice of India, Honourable Mr. Justice S.A. Bobde since 2019. He has a keen interest in environmental law, which he intends to practise in the courts of India.

Bharat Patel is the Senior Program Manager at the Centre for Policy Research–Namati Environmental Justice Program, New Delhi, and General Secretary for the Machimar Adhikar Sangharsh Sangathan (MASS), Bhuvneshwar, Kutch, Gujarat. Dr Patel obtained his Ph.D. from the Gujarat Vidhyapith and has been working with fishers' cooperatives in Kutch for more than a decade. His recent work is focused on mobilizing fishers and coastal communities against the rapid industrial development of the Gujarat coastal region.

Kirit Patel is an Associate Professor and the Program Chair of International Development Studies at Menno Simons College, affiliated with the University of Winnipeg and Canadian Mennonite University. As an academic, policy analyst and development professional, Dr Patel's teaching and research focus on environmental and social justice, sustainable food systems and nutrition security, agrarian change and rural–urban migration, agrobiodiversity conservation, Indigenous knowledge systems and the governance of common property resources in South Asia.

Suhaas Putta is an alumnus of Jindal Global Law School. His main interests are international environmental law and international criminal law. He is currently a Graduate Research Immersion Programme Scholar at O.P. Jindal Global University and is researching the viability of trying environmental crimes in international tribunals.

Armin Rosencranz is the Dean of the Jindal School of Environment and Sustainability. He was a faculty member at Stanford University and taught environmental and natural resources policy and law for 20 years. His courses were cross-listed in ten different departments, including history, political science, human biology and earth systems. He received three student-nominated awards for teaching excellence, including "teacher of the year."

R. Seenivasan is trained in agricultural engineering and rural management and has a Ph.D. in law from the University of Westminster. He has worked as a development professional, focusing on rehabilitating India's traditional irrigation tanks and networks, participatory water management, and promoting credit networks and unions supporting rural women. His academic interests revolve around his development work as well as understanding the social conflicts created by the law and technology.

A. John Sinclair is a professor and the Director of the Natural Resources Institute, University of Manitoba. His main research interest focuses on governance and learning as they relate to resource and environmental decision-making. Applied research in these areas takes him to various locations in Canada, Africa and Asia. He has a very active research programme spanning over 25 years in relation to environmental assessment law and policy. Through current research grants he is considering the role of learning in meaningful participation, best approaches for incorporating climate change considerations in assessment and the elements of next-generation assessment.

Cassandra Szabo is a social researcher who has been involved in multiple research projects in northern Canada and one project in India, all focusing on sustainable development and, specifically, on socially inclusive development. Her interests centre on understanding how communities can foster growth without suffering negative social, environmental or health outcomes.

Neelotpalam Tiwari is a 2019 law graduate from Gujarat National Law University, India. A lawyer by profession, he also undertakes independent research activities in various areas, including but not limited to environmental issues and dispute settlement. He has interned with several regional organizations in order to better understand the practical application of the law.

Acknowledgements

We gratefully acknowledge the financial assistance of the Social Sciences and Humanities Research Council of Canada (Insight Development Grant 430-2016-01025), which supported student training and the direct costs of research in India. We also thank the Queen Elizabeth II Diamond Jubilee Scholarships programme, which funded student internships in India and provided assistance for Indian advanced scholars to take up positions in Canada. Our respective universities – the University of Winnipeg; Menno Simons College, Canadian Mennonite University; and Gujarat National Law University – also provided considerable financial and in-kind assistance to support research, travel, networking and knowledge exchange. Similarly, we wish to thank our partner organizations in Canada (Manitoba Eco-Network, International Institute for Sustainable Development) and India (Centre for Policy Research, Foundation for Ecological Security, Machimar Adhikar Sangharsh Sangathan, Sadbhavna Hospital Trust, Snehakunja Trust) for their in-kind and research assistance. Special thanks also go to Maureen Epp for her excellent editorial assistance and to Weldon Hiebert for his top-notch cartographic skills. Finally, we are deeply grateful to our many research participants who unfailingly were generous with their time and hospitality and often shared profoundly insightful observations and personal concerns about their livelihoods, families and communities.

Alan P. Diduck
Kirit Patel
Aruna Kumar Malik

Part I

Introduction

1 Environmental justice in India

Context, issues and framework

Kirit Patel, Alan P. Diduck and Aruna Kumar Malik

Introduction: historical institutional and policy context

The original Constitution of India passed by the Constituent Assembly on 26 November 1949 did not make any reference to environmental protection, rights or justice; nevertheless, India has made great strides by introducing an array of legislation, policies and programmes to protect the environment (Dwivedi, 1997). Recognizing its commitment made to the international community at the 1972 United Nations Conference on the Environment in Stockholm, India made constitutional amendments to recognize the duties of citizens to protect the environment and subsequently passed several acts to protect water, air, wildlife, forests and biodiversity. Subsequently, it established several institutions, including the Ministry of Environment and Forest in 1985, for implementing environmental legislation, policies and programmes (Bhat, 2010; Sahasranaman, 2012). These and other environmental policy initiatives faced new and formidable challenges after India opened its economy in 1991. The globalization and structural adjustment policies introduced in the early 1990s promoted intensive industrial growth, privatization, foreign direct investment and consumption of goods supplied by international markets. Recognition of these new challenges prompted creation of new institutions and policy measures such as the Environment Impact Assessment (EIA) Notification of 1994 and the National Environment Appellate Authority in 1997.

At the end of the 20th century, India became confident of sustaining a higher rate of economic growth driven by infrastructure development, industrial production and international trade. An economic surge between 2008 and 2018, marked by an average annual GDP increase of 7 per cent (IMF, 2019), was clouded by income inequality, rising poverty and widespread environmental degradation. In 2011, approximately 22.5 per cent of India's population was living in poverty (United Nation Development Program, 2020; World Bank, 2020), and the prevalence of moderate and severe food insecurity (2017–2019) sits at about 31.6 per cent (Bansal, 2020; FAO et al., 2020). The courts received numerous environmental complaints from communities or local activists (Kukreti, 2019) and thus became important actors

DOI: 10.4324/9781003141228-1

in protecting the environment, owing to the failure of government institutions responsible for enforcement of environmental legislation such as the EIA Notification. As this was occurring, the Indian judiciary, known for its creative interpretation of various provisions of the Constitution to protect the rights of marginalized sections of society, introduced several institutional innovations to adjudicate environmental cases (Patel and Dey, 2013).

Realizing the importance of technical expertise in environmental cases, the Supreme Court of India and several state-level high courts established green benches to adjudicate environmental disputes. In spite of disparities in structure, context of origin, availability of institutional resources and environmental policies, each of these green benches has made significant contributions to environmental jurisprudence in India. However, considering the importance of institutional consistency in environmental jurisprudence, the Law Commission of India (2003) recommended establishing formal specialized environmental courts. After lengthy parliamentary processes, the National Green Tribunal (NGT) Act was passed in 2010. The Indian judiciary, thus, entered a new era of environmental jurisprudence with the creation of a specialized judicial forum with a mandate to effectively and expeditiously dispose of cases relating to the environment (Dutta and Purohit, 2015; Gill 2010, 2017).

These reforms brought positive changes to the traditional functioning of courts in interpreting environmental law and using scientific expertise in resolving litigation. They also instilled a sense of societal responsibility in the judiciary (Mathur, 2007). However, scholarly opinion on the impact of such reforms on society and the poor are divided (e.g. Sahu, 2014; Suresh and Narrain, 2019). Furthermore, the efficacy of the reforms in advancing the cause of environmental justice requires close examination. Environmental justice is complicated and multifaceted, as explained later in the section on theoretical orientation, but at its heart is the equitable distribution of environmental risks and benefits. This basic principle suggests that any examination of judicial reforms and environmental justice also requires careful consideration of the environmental policy context, including requirements for EIA (e.g. Ewall, 2012) and provisions for enforcement of regulatory requirements. Moreover, such investigations obviously warrant a critical examination of the influence of the neoliberal state on the functioning of courts and institutions responsible for implementing environmental policies and rendering justice. In the context of environmental litigation in India, the state often has multiple and conflicting roles as proprietor, petitioner, regulator, arbitrator and polluter. It is, therefore, imperative to examine how the policy regime and institutions responsible for ensuring compliance, including the constitutional courts and tribunals, are influenced by the political economy and the development ideology of the state.

This book, thus, examines the effectiveness of environmental policies and associated regulatory and compliance institutions, including the judiciary and the NGT, in rendering environmental justice for marginalized communities while ensuring protection of the environment. It also analyses the influence of the neoliberal state and its political economies on the governance and outcomes of these environmental policies and institutions. Finally, the book presents numerous suggestions for advancing environmental justice through legal, policy and institutional reform.

Policy and institutional context and reforms for environmental justice

Environmental policies and associated institutions: reforms and repercussion

As noted earlier, following the 1972 Stockholm Declaration, India undertook a series of initiatives to bolster its environmental policy regime and institutional framework (Bhat, 2010; Sahasranaman, 2012). Along with the constitutional reforms mentioned above, an extensive array of legislation and associated regulatory agencies was established, as were national and state policies and strategies that provided guiding principles (OECD, 2006; Dwivedi, 1997). Among the key national laws that were first adopted were statutes focused on particular environmental components (i.e., air, water, land and biological resources): the Wildlife Protection Act 1972; the Water (Prevention and Control of Pollution) Act 1974; the Water (Prevention and Control of Pollution) Cess Act 1977; the Forest (Conservation) Act 1980 and the Air (Prevention and Control of Pollution) Act 1981. This legislation also established various agencies with substantial regulatory authority, including standard-setting, permit/consent-granting, monitoring and enforcement responsibilities. For example, the Water (Prevention and Control of Pollution) Act and the Air (Prevention and Control of Pollution) Act vested authority in the Central and State Pollution Control Boards to enforce emission and effluent standards for industries discharging pollutants into the air and water.

Following the Bhopal disaster of 1984, the central government extended the policy and institutional framework by adopting integrated legislation – the Environment (Protection) Act 1986 – that created an overarching national environmental protection regime and established the Ministry of Environment and Forest, now the Ministry of Environment, Forest and Climate Change (MoEFCC) (OECD, 2006). In 1994, the act was used to advance a more anticipatory and preventative environmental agenda by providing the basis for India's first law-based EIA regime, later superseded by the Environment Impact Assessment Notification 2006. EIA law and practice have evolved in India over the years, with a recent comprehensive appraisal by Jha-Thakur and Khosravi (2021) revealing that progress has been made with respect to openness and public character, while few advancements have been seen with regard to completeness, objectivity and verifiability. Openness and public character encompass features such as community engagement, use of local knowledge and transparency. Completeness deals with questions such as what undertakings are subject to assessment, the range of impacts that will be considered and whether effective techniques and methodologies will be employed. Objectivity is concerned with the credibility and neutrality of the EIA process, including the expertise and impartiality of the people who prepare and review EIA reports. Finally, verifiability pertains to the effectiveness and external oversight of the EIA process, including the accuracy of impact predictions and the efficacy of mitigation measures, as well as whether the EIA results influence the decision-making process.

The fact that few advancements have been made with respect to completeness, objectivity and verifiability helps explain the extent to which EIA process and practice have attracted considerable scholarly criticism (e.g. the review in Jha-Thakur and Khosravi, 2021) and have provided substantial grounds for court challenges (e.g. the numerous legal cases cited in Chapters 5 and 6). Additionally, the government's executive branches initially lacked the resources, capacity and political will to aggressively enforce the EIA Notification and other legislation noted above, such as the Water (Prevention and Control of Pollution) Act and the Air (Prevention and Control of Pollution) Act (Dwivedi, 1997; OECD, 2006). India's early policy and institutional mechanisms for advancing environmental justice fell far short of the mark, and the courts, therefore, became important actors in protecting and advancing environmental justice (Dembowski, 2003; Gill, 2017).

Institutional change in the judiciary for environmental justice

The Supreme Court of India's innovative interpretation of the Constitution to relax requirements for locus standi in the early 1980s significantly advanced the cause of social and environmental justice (Bhagwati, 1985; Cassels, 1989; Peiris, 1991). This created a unique legal framework for public interest litigation (PIL) wherein civil society groups and individual activists could approach the judiciary, even as a non-affected third party, to seek suitable legal remedies for poor and marginalized sectors of society. The provision of PIL has been criticized for, among other things, enabling the exercise of judicial power unconstrained by procedural norms and processual limits (Bhuwania, 2017; Gauri, 2009). Nevertheless, PIL gave rise to a new era of activism in India: nongovernmental organizations, farmers' associations, civil society organizations and activists were empowered in their efforts to protect the environmental and social rights of vulnerable classes (Bhushan, 2004; Curmally, 2002). The significance of the relaxation of locus standi and ensuing PIL was evident from the high volume of cases that sought mandamus action from the court that would compel the government to protect environmental and social rights (Gauri, 2009; Sahu, 2008).

When the NGT began operations in 2011, India joined a group of 41 nation states that had created specialized courts or tribunals for adjudicating environmental cases (Pring and Pring, 2016). The NGT was meant to offer a one-stop shop for effective resolution of environmental disputes and thus to reduce the probability of irreversible environmental damage resulting from delayed and scientifically unsound judgements of conventional courts. The NGT meets several best practices identified by Pring and Pring (2009) for effective environmental jurisprudence. It includes an open rule of standing and has embraced several international principles laid out in the 1992 Rio Declaration, such as the precautionary principle, the polluter pays principle and various dimensions of sustainable development (Patel and Dey, 2013; Gill, 2010, 2019). Furthermore, the NGT (Article 15(1)) is empowered to order relief or compensation for victims of pollution or persons who are aggrieved by the environmental damage arising from decisions or orders made by the institutions that provide environmental clearances. However, the NGT has only a limited ambit of litigation arising from fewer than seven pieces of national legislation. A vast number of environmental laws at the provincial level are outside the NGT's mandate.

Although the long-term impacts of the NGT have yet to be determined, some observers have commented favourably on the environmental implications of the decisions that have been rendered (e.g. Patra and Krishna, 2015; Rengarajan, Palaniyappan, Ramachandran and Ramachandran, 2018). Others have been more critical, highlighting important gaps and potential shortcomings in the NGT's institutional arrangements. For example, some authors have noted the historical failure in India of quasi-judicial bodies set up outside the conventional court system resulting from political and administrative interference (Kohli, 2011; Nambiar, 2012; Rosencranz and Sahu, 2009; Rosencranz, Sahu and Raghuvanshi, 2009; Sharma, 2008). In fact, the NGT, being a tribunal or specialized court, has been vulnerable to withdrawal of financial, administrative and political support from the national government, and this has undermined its functioning and threatened its independence (Ghosh, Sanyal, Chandrashekar and Sekhar, 2018; Sahu, 2018). Gill (2020) has gone so far as to argue that the MoEFCC has, over the years, been among the primary restraining forces on the full and efficacious functioning of the NGT. Other restraints have come from political and economic interests, selected policy and legislative interventions and various regulatory agencies.

One of the practical concerns for rendering environment justice through the NGT is access for the poor. The NGT serves in only five locations throughout the country, with the main branch in Delhi and circuit branches in Kolkata, Pune, Bhopal and Chennai. The Pune branch, serving several states in western India, remained closed for two years from 2018 to 2020 owing to appointment-related issues (Hindustan Times, 2020). The lower level civil courts, located at each block level, are undoubtedly the first venue of action for many local communities to seek environmental redress, as they are within travelling distance to rural villages. However, under the NGT Act, civil courts will no longer have jurisdiction in matters relating to the environment (Article 29(1) and 29(2)). Furthermore, civil courts conduct their business primarily in local languages and lawyer's fees in those local courts are relatively low. Transferring all environmental cases to principal or circuit branches of the NGT is disadvantageous for the poor considering cost of travel, language barriers and higher legal fees. Moreover, the tensions between state-level high courts and the regional NGT arising from jurisdictional and legislative overlaps have yet to be seriously addressed. Thus, the importance of high courts continues for environmental justice even after the establishment of the NGT.

The neoliberal state and the courts: consequences for the poor

Protecting and advancing environmental justice in India requires more than attending to environmental policies and judicial institutions. Careful consideration must also be given to the broader economic policies and political narratives that set the bounds for and otherwise constrain environmental policy and institutional reform. In India's case, economic liberalization has been a powerful force. Since liberalization began in the early 1990s, India has become a perfect example of Kuznet's inverted-U curve, where income inequality and ecological degradation have continued to increase along with economic growth (D'Souza, 2005). The ramifications of ecological degradation are particularly harsh for the poor, who are directly exposed to environmentally hazardous living and working conditions. The efficacy of institutions rendering environmental justice should be measured by how effectively they protect the constitutional rights and the well-being of poor and marginalized sections of society. The creation of PIL and other institutional reforms for resolving environmental disputes is undoubtedly a credible indicator of the judiciary's intention to address poverty, social exclusion, income inequality and power imbalances (Patel and Dey, 2013). However, the pro-poor reputation of the judiciary has come under question in the neoliberal era. In a study examining PIL filed in the Delhi Supreme Court, Gauri (2009) found success rates for legal actions seeking the protection of fundamental rights were significantly higher in recent years when the claimants were from advantaged socio-economic groups. This trend indicates social reversal from the original objective of introducing PIL by the judiciary.

Other scholars have also argued that judicial institutions have taken a neoliberal turn in deciding case priorities and in legal analysis (Suresh and Narrain, 2019) and they tend to negate the interests of the poor (Aggarwal, 2005; Shekar and Thomas, 2015). For instance, in the case of *Shree Mahuva Bandhara Khetiwadi Pariyavaran Bachav Samittee v Union of India* (2010), the court acknowledged that the disputed cement factory was located in a wetland, adversely impacting the coastal environment and livelihood of small farmers and pastoralists in the area. However, in rejecting the local community's plea to stop construction of the cement factory, the high court judge wrote, "Undoubtedly, progress and development would require [the] construction of factories. Such factories would inevitably increase production and generate employment in [the] rural area." This type of judicial perspective reflects and endorses the neoliberal economic growth narrative of the central and many state governments in India (Dilay, Diduck and Patel, 2020).

Political ecologists argue that the environment is not a neutral entity; it represents different material and cultural significance for people depending on their socio-economic class, ethnicity and gender (Véron, 2006). The cases filed by environmental organizations and activists primarily reflect the priorities and values of the urban middle class (Bhan, 2009). For instance, a series of judicial orders by the Supreme Court to close small polluting factories and remove squatter settlements from public spaces in Delhi had profound negative impacts on the livelihoods of more than two million urban poor. Baviskar (2003) and other scholars (Bhan, 2009; Gonsalvez and Sahu, 2010; Rajamani, 2007) labelled such environmental litigation as bourgeois environmentalism aided by English-speaking, urban middle-class activists, journalists and judges.

In the context of environmental justice, the poor have often become victims of the same policies and institutions that were originally introduced to liberate them from poverty, pollution and a skewed power structure. The courts are becoming increasingly reluctant to admit cases against state-sponsored infrastructure and industrial development projects. Bhushan (2009) argues that whenever environmental protection comes into conflict with private-sector interests, the courts tend to interpret the interest of the private sector as "economic development" in societal interests. In these circumstances, courts often subordinate the environmental cause and side with the private sector supported by the neoliberal state. Furthermore, the inaction of environmental courts also punishes the poor. For instance, the December 2015 flood in Chennai, causing more than 470 deaths and property devastation affecting 1.4 million poor households, is blamed on a failure to protect water infrastructure from massive urban projects and sprawl (Radhakrishnan, 2015). Several court cases were filed for protecting water resources and drainage infrastructure in Chennai in years prior to the flood, but they met with limited or no success (Seenivasan 2014, 2016). If courts had taken any preventative action in these cases, it would have been considered against the interest of urban development.

Furthermore, legislative branches in India conveniently seek intervention from the judicial branch to make unpopular or anti-poor decisions that may backfire in the next election. For instance, if slum dwellers have to be removed to build residential apartments for the middle class or commercial property, the government may hesitate to make such a decision for fear of democratic backlash from the poor. The courts, being an authority of law and an indirect ally of the government's neoliberal values and development agenda, are viewed as the best instrument to reinforce the interests of the higher and middle classes. Gauri (2009) attributes the urban middle-class bias of courts to class structure that is ingrained in the Indian judiciary.

Theoretical orientation

The conceptual framework at the heart of this book draws upon theoretical insights of law and development scholars like Trubek (1980) and others who have questioned traditional meanings of environmental justice based simply on principles of fair treatment and equity in distribution of environmental risks and benefits (Williams and Mawdsley, 2006). Critical analysis of environmental movements indicates three important attributes of environmental justice: recognition of the diversity of participants and experiences in affected communities (recognitional justice); opportunities for participation in the political processes that create and manage environmental policy (procedural justice); and equity in the distribution of environmental risk and benefits (distributive justice) (Coolsaet, 2020; Schlosberg, 2004; Whyte, 2018). Additionally, practitioners and scholars have recently highlighted the salience of a fourth attribute: the extent to which negative environmental and social impacts are remedied (restorative justice) (Motupalli, 2018; Raphael, 2019).

Moreover, drawing upon Amartya Sen's and Martha Nussbaum's capabilities approach, Schlosberg and Carruthers (2010) argue that environmental justice should also be evaluated on how so-called fair distributions obtained through environmental litigation affect the ultimate well-being, participation and functioning of people's lives. This necessitates understanding how the intersection between neoliberal government and markets empowers or silences civil society (which expresses alternatives to state and market values) and poor citizens (the shock absorbers of society) who are forced to stay and suffer or flee from environmental degradation (Thompson, 2008). In addition, the procedural dimension of the environmental justice framework encompasses the three pillars of environmental democracy established in the Aarhus Convention (United Nations Economic Commission for Europe, 2019): access to information, access to participation in decision-making and access to justice. With respect to access to justice, Pring and Pring (2009) suggest that efficacy be judged at three different stages of environmental litigation. The first encompasses getting to and through the courthouse door, and having the information and knowledge, the standing, the legal and technical support, and the ability to take on the risks of litigation. The middle stage involves access within the environmental courts to proceedings that are fair, efficient and affordable. The end stage includes access to enforcement remedies and tools that can implement the court's decision.

The four interlinked dimensions of environmental justice (see Figure 1.1) are founded on basic attributes of good governance, such as broad-based citizen participation, respect for the rule of law, respect for cultural values, effectiveness and efficiency, accountability, transparency and responsiveness (United Nations Economic and Social Commission for Asia and the Pacific, 2008). The framework is also informed by the goals of sustainable development, such as inter- and intragenerational equity (World Commission on Environment and Development, 1987) and the United Nations Sustainable

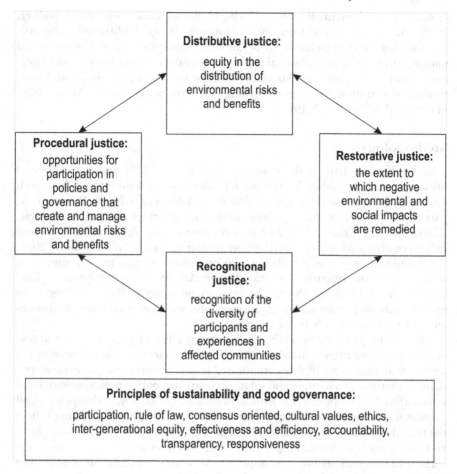

Figure 1.1 The four attributes of environmental justice, founded on the principles of good governance.

Development Goal #16. This goal is to "promote peaceful and inclusive societies for sustainable development, provide access to justice for all and build effective, accountable and inclusive institutions at all levels" (United Nations Department of Economic and Social Affairs, 2020).

The conceptual focus of this book aligns with the historical foundations of environmental justice, namely litigation, advocacy and regulation in response to racial and income disparities. It also fits with how environmental justice analysis has been extended to a wider array of inequities and affected communities, intersectional analysis and concern for equitable access to environmental benefits (Coolsaet, 2020; Holifield, Chakraborty and Walker, 2018; Malin, Ryder and Lyra, 2019; Raphael, 2019). Additionally, the policy and institutional orientation of the book complements the many broad national and transnational social movements concerned with aspects of environmental justice, such as fair trade, climate justice, water justice and protection of common-pool resources (e.g. Environmental Justice Atlas, 2020; Harley and Scandrett, 2019).

Methodology

This book stems from a three-year project (2017–2020) on environmental justice funded primarily by the Social Sciences and Humanities Research Council of Canada and the Queen Elizabeth Advanced Scholars Program. As part of the project, we three editors undertook primary research in Gujarat, Karnataka and Tamil Nadu and met numerous colleagues from India and other countries who were involved in similar research. In December 2018, we invited some of these colleagues to contribute to this book project and at the same time organized an authors' workshop at Gujarat National Law University on 14 March 2019. The primary purposes of the workshop were to seek common understandings of environmental justice concerns and issues and discuss potential chapter content.

The central problematic of the book, namely advancing environmental justice for poor and marginalized sectors of Indian society, is inherently interdisciplinary. It addresses a highly complicated problem and encompasses concepts, themes, methods, challenges and solutions from law, public policy, governance, international development, environment studies, geography, anthropology and economics. Some chapters deepen the book's interdisciplinarity, taking it near the transdisciplinary realm. In some of the research reported in the book, local and Indigenous understandings, values and interests are central to how the authors define judicial impacts, inquire into solutions to adverse impacts and formulate recommendations for legal, policy and governance solutions.

Overall, the book draws upon theories and understanding from three applied fields: (1) theories and approaches related to rights-based development of poor and marginalized people; (2) theories and practice of participatory conservation and environmental governance; and (3) understanding of legal innovations, processes and the judiciary in the Indian context. Some chapters in the book reflect an explicit critical theory orientation, while others adopt paradigms of legal research that tend to focus on constitutional jurisdiction, interpretation of common and statute law and jurisprudence. Still other chapters emphasize the impacts of development or judicial decisions on the poor and the institutional contexts in which those decisions occur.

With respect to methods, some chapters rely heavily on legal and policy analysis of legislation, policies and judicial decisions. Others use ideas and field research methods from anthropology, development studies, human geography and sociology. All deal with one or more aspects of the environmental justice framework described above. Overall, the orientation to knowledge production is critical, usually constructivist and often interdisciplinary or transdisciplinary. Individually and collectively, the chapters examine a range of environmental and natural resource issues, such as groundwater, invasive species and air and water pollution, along with a diversity of state-sponsored, commissioned or sanctioned industrial projects spanning energy production (small hydro dams, a coal-fired power plant), mining (limestone, sand, copper and bauxite) and infrastructure (ports and highways). Chapter-specific research methods, especially where based on empirical case studies, are described in each chapter.

Contributions and organization of the book

The policy and institutional orientation of the book has as its primary emphasis an interrogation of legislative, judicial and policy initiatives, such as PIL, green benches, the NGT and environmental impact assessment. Examining these initiatives through the lens of the four dimensions of environmental justice sheds light on feasible options for policy and institutional reform, including with respect to effectiveness, equity and various facets of sustainable development. These suggestions for reform are sprinkled throughout the book and reinforced in the concluding chapter.

The book also reveals whether state high courts and the NGT, in the process of enforcing the rule of law and protecting the environment, have become instruments that reinforce existing socio-economic inequality by externalizing environmental harm and the costs of environmental protection to marginalized sections of society. This analysis involves examining the impact of these judicial forums for their consistency (or lack thereof) in using widely recognized principles regarding impact assessment, sustainable development, public participation and human rights. In terms of the role of the judiciary in a liberal democratic state, the book reveals whether courts empower marginalized sections of society to actively participate in governance and seek environmental protection. In doing so, the book helps us understand the relationships between litigation outcomes and development ideology in the neoliberal state.

Overall, the book provides a unique perspective on environmental justice because of its policy and institutional orientation along with a consistent emphasis on social justice. Such a perspective helps us understand and respond to the neoliberal state's ability to empower or suppress civil society, which often forces marginalized sectors of society to either tolerate or escape environmental degradation (Thompson, 2008). Emphasizing policy and institutional development also contributes to critical thinking on rights-based approaches that seek empowerment of the poor through judicial and other governance institutions.

The book is organized into four parts. Following this introductory section, Part II establishes important components of the policy, economic and institutional context of recent struggles for environmental justice in India. In Chapter 2, Mark Baker argues that small hydro policy in Himachal Pradesh exemplifies hallmark characteristics of neoliberalization and negatively affects household reproduction and the environments on which rural communities depend. He describes how local resistance has emerged in opposition to these environmental injustices in order to defend rural lifeways, and he offers two promising community-oriented institutional exceptions to neoliberal small hydro policy regimes.

Arpitha Kodiveri, in Chapter 3, describes a conflict between Adivasi (Indigenous) communities and a multinational company attempting to acquire land for a bauxite mine in Odisha. She situates the conflict in the historical development of India's current neoliberal policy regime and argues that the current emphasis on ease of doing business has been accompanied by legislative attempts to dilute the right of consent under the Forest Rights Act (2006). She further argues that forest-dwelling communities are unable to participate adequately in decisions about acquisition of forest land, resulting in forceful acquisition of these lands and resources. Finally, Kodiveri describes the potential and limits of jurisdictional leapfrogging, in which international institutions are petitioned to intervene in domestic affairs.

In Chapter 4, Gayathri Naik adopts a water justice perspective and reveals injustices with respect to groundwater because of common law principles that confine the benefits of groundwater access to landowners, while the burdens of scarcity are shared more widely. In doing so, she provides an introduction to an important aspect of the constitutional and jurisprudential context for environmental justice, namely the right to life guaranteed by Article 21 of the Constitution of India. The chapter concludes with a call for improved legislative support for procedural justice, including rights to information, public participation and access to justice.

In Chapter 5, Armin Rosencranz and colleagues analyse selected legal cases pertaining to impact assessment, rehabilitation, compensation and related policy measures to appraise the NGT's response to the cause of two marginalized communities: tribal peoples and fisherfolk. They argue that although not perfect, the NGT's efforts to balance economic and human development with environmental protection are commendable. They conclude that to advance environmental justice, it is important for the tribunal to stay its pragmatic course, all the while ensuring that its decisions are well reasoned and consistent.

Himanshu Pabreja and Neelotpalam Tiwari summarize the environmental clearance process in Chapter 6 and take an in-depth look at a key aspect of the EIA Notification 2006, namely the expert appraisal phase. They analyse the roles, composition and functioning of expert appraisal committees by reviewing pertinent legislative provisions as well as judicial and NGT decisions from 2010 to 2019. The analysis raises concerns over the appraisal process and the judicial oversight of the process. Among their suggestions for reform, the authors argue for improved vetting of EIA documents to ensure adequacy, accuracy and veracity, and stricter control of project proponents and regulatory authorities who are derelict in their statutory duties.

Part III of the book features eight case studies based on empirical field research that covers a selection of geographic, environmental and natural resource settings. The cases reveal different combinations of recognitional, procedural, distributive and restorative injustices, a sampling of environmental justice success stories and further evidence regarding context variables described in Part II. In Chapter 7, Alan Diduck and colleagues revisit small hydro policy in Himachal Pradesh and supplement Baker's political economy analysis with a close look at five hydro projects in a single district. By focusing on specific environmental approval and clearance processes and oppositional movements, the chapter offers suggestions for policy reforms, including greater recognition of the knowledge and acumen of local mountain people and more effective integration of small hydro planning with impact assessment and disaster risk reduction.

Chapters 8, 9 and 10 focus on industrial developments in Gujarat. In the first of these, Ariane Dilay examines a limestone mine in Bhavnagar District and describes how inadequate public engagement and the threat of significant adverse local impacts led to conflict, lengthy litigation and widespread social opposition. Among other lessons, the case highlights a strong need for improved information sharing and mandatory early public engagement in the impact assessment process. In Chapter 9, Bryce Gallant's gender-focused analysis of a major cement factory in Bhavnagar District reveals similar problems, conflicts and lessons. She also argues that impact assessment could play a key role in achieving environmental justice if it were implemented intersectionally. Gender-based analysis, increasingly used in impact assessments worldwide, is a promising starting point for achieving such a goal. Chapter 10, by Avery Letkemann and colleagues, presents a case study involving a coal-fired power plant in Kutch District. As with the other chapters in this group, the case highlights recognitional, procedural and distributional inequities, legal challenges and protest movements. The unique aspect of this case, however, lies in the efforts made by local people to seek remedies from one of the financiers of the power plant, the International Finance Corporation of the World Bank Group, their receiving support from EarthRights International and launching a suit again the International Finance Corporation in the American courts. At the time of writing, these efforts have not borne fruit, but they suggest new and creative ways for local communities to advance their interests.

Chapters 11 and 12 present case studies from Karnataka. The first of these, by Cassandra Szabo and Mahabaleshwar Hegde, examines sand mining in the Swarna River. The case involves a petition by local residents to the NGT to overturn permits granted by the district administration. The petitioners were successful but they faced several challenges in gaining justice, such as securing legal and technical support and accessing enforcement remedies and tools to effect the tribunal's decision. Mahabaleshwar Hegde and colleagues scrutinize the environmental clearance granted for a major expansion of the Karwar port in Chapter 12. They describe serious flaws in the impact assessment process, including concealment of vital information, failure to recognize comprehensive impacts and failure to suggest restorative and preventative measures. For the authors, these problems raise questions about the competence of the scientific and environmental regulatory agencies involved. The chapter also reveals, as others in this volume do, how seeking justice through judicial means can place a significant burden on rural communities.

Chapters 13 and 14 focus on two longstanding problems in Tamil Nadu. In the former, Aruna Malik and colleagues appraise environmental assessment, clearance, enforcement and compliance mechanisms related to the approval and operation of the controversial Sterlite Copper smelter plant in Thoothukudi. The case reveals significant regulatory and licensing violations by the company, serious gaps in enforcement and compliance on the part of environmental regulatory agencies and a need for fundamental legal and policy reform. As a result of the plant's ongoing operation, environmental conditions, community health and livelihood security steadily deteriorated and trust in the state, governance institutions and law became increasingly fragile. Ultimately, protests demanding the plant's closure turned tragic when in May 2018 at least 14 people were killed in police fire. In Chapter 14, R. Seenivasan problematizes a judicial remedy that had been granted following a public interest petition to address the spread of *Prosopis juliflora*, an invasive plant species introduced in the colonial era. The author argues that a series of far-reaching orders made by the Madras High Court for removal of the plant from land and water in one-third of the state is a dire example of judicial overreach. The orders aimed to protect the environment but did so at the expense of rural livelihoods and property rights. The approach of the court produced recognitional and procedural injustices, resulting in perverse distributive and restorative outcomes.

References

Aggarwal, S. (2005) *The Public Interest Litigation Hoax in India: Its Adverse Impact on the Poor and Working Class*. South Asia Citizens Web. Available at: http://www.sacw.net/article11105.html (Accessed: 5 January 2021).

Bansal, V. (2020) "More evidence of India's food insecurity," *The Hindu*, 24 August.

Baviskar, A. (2003) "Between violence and desire: Space, power, and identity in the making of metropolitan Delhi," *International Social Science Journal*, 55, 89–98.

Bhagwati, P.N. (1985) "Judicial activism and public interest litigation," *Columbia Journal of Transnational Law*, 23, 561–578.

Bhan, G. (2009) "'This is no longer the city I once knew': Evictions, the urban poor and the right to the city in millennial Delhi," *Environment and Urbanization*, 21(1), 127–142.

Bhat, S. (2010) *Natural Resources Conservation Law*. New Delhi: Sage.

Bhushan, P. (2004) "Supreme Court and PIL: Changing perspectives under liberalization," *Economic and Political Weekly*, 39(18), 1770–1774.

Bhushan, P. (2009) "Misplaced priorities and class bias of the judiciary," *Economic and Political Weekly*, 44(14), 32–37.

Bhuwania, A. (2017) *Courting the People: Public Interest Litigation in Post-Emergency India.* Cambridge: Cambridge University Press.

Cassels, J. (1989) "Judicial activism and public interest litigation in India: Attempting the impossible?" *The American Journal of Comparative Law*, 37(3), 495–519.

Coolsaet, B. (ed.) (2020) *Environmental Justice.* London: Routledge.

Curmally, A. (2002) "Environment and rehabilitation" in Morris, S. and Shekar, R. (eds.) *Indian Infrastructure Report 2002: Governance Issues for Commercialization.* New Delhi: Oxford University Press, 96–108.

Dembowski, H. (2003) *Taking the State to Court: Public Litigation and the Public Sphere in Metropolitan India.* Oxford: Oxford University Press.

Dilay, A., Diduck, A.P. and Patel, K. (2020) "Environmental justice in India: A case study of environmental impact assessment, community engagement and public interest litigation," *Impact Assessment and Project Appraisal*, 38(1), 16–27.

D'Souza, R. (2005) "The 'Third World' and socio-legal studies: Neo-liberalism and lessons from India's legal innovations," *Social and Legal Studies*, 14(4), 487–513.

Dutta, R. and Purohit, S. (2015) *Commentary on the National Green Tribunal Act, 2010.* New Delhi: Universal Law Publishing.

Dwivedi, O.P. (1997) "Green justice: An overview of environmental laws and regulations" in Dwivedi, O.P. (ed.) *India's Environmental Policies, Programmes and Stewardship.* New York, NY: St. Martin's Press, 79–110.

Environmental Justice Atlas (2020) Environmental Justice Atlas. Institute of Environmental Science and Technology, Universitat Autonoma de Barcelona. Available at: https://ejatlas.org (Accessed: 15 October 2020).

Ewall, M. (2012) "Legal tools and environmental equity vs environmental justice," *Sustainable Development Law and Policy*, 13(1), 4–13.

FAO, IFAD, UNICEF, WFP and WHO (2020). *The State of Food Security and Nutrition in the World 2020: Transforming Food Systems for Affordable Healthy Diets.* Rome: FAO. Available at https://doi.org/10.4060/ca9692en (Accessed: 16 February 2021).

Gauri, V. (2009) *Public Interest Litigation in India: Overreaching or Underachieving?* Policy Research Working Paper 5109. Washington, DC: The World Bank, Development Research Group, Human Development and Public Services Team.

Ghosh, A., Sanyal, D., Chandrashekar, R. and Sekhar, R. (2018) *Reforming the Tribunals Framework in India: An Interim Report.* New Delhi: Vidhi Centre for Legal Policy.

Gill, G.N. (2010) "A green tribunal for India," *Journal of Environmental Law*, 22(3), 461–474.

Gill, G.N. (2017) *Environmental Justice in India: The National Green Tribunal.* Abingdon, Oxon: Routledge.

Gill, G.N. (2019) "Precautionary principle, its interpretation and application by the Indian judiciary: 'When I use a word it means just what I choose it to mean – Neither more nor less' Humpty Dumpty," *Environmental Law Review*, 21(4), 292–308.

Gill, G.N. (2020) "Mapping the power struggles of the National Green Tribunal of India: The rise and fall?" *Asian Journal of Law and Society*, 7(1), 85–126.

Gonsalvez, C. and Sahu, G. (2010) "Litigating against corporations for human rights," *Economic and Political Weekly*, 46(14), 18–21.

Harley, A. and Scandrett, E. (2019) *Environmental Justice, Popular Struggle and Community Development.* Bristol: Policy Press.

Hindustan Times (2020) "Pune-based western zone bench of NGT to start operation after two years in Feb 2020," *Hindustan Times*, 2 January. Available at: https://www.hindustantimes.com/pune-news/pune-based-western-zone-bench-of-ngt-to-start-operations-after-two-years-in-feb-2020/story-fMG9PZODprG37a-zlTxGQcI.html (Accessed: 16 February 2021).

Holifield, R., Chakraborty, J. and Walker, G. (eds.) (2018) *The Routledge Handbook of Environmental Justice*. London: Routledge.

IMF (International Monetary Fund) (2019) *IMF Country Information: India*. International Monetary Fund. Available at: https://www.imf.org/en/Countries/IND#countrydata (Accessed: 12 January 2019).

Jha-Thakur, U. and Khosravi, F. (2021) "Beyond 25 years of EIA in India: Retrospection and way forward," *Environmental Impact Assessment Review*, 87, 106533. Available at: https://doi.org/10.1016/j.eiar.2020.106533.

Kohli, K. (2011) "A functional Green Tribunal," *India Water Portal*, 12 July. Available at: https://www.indiawaterportal.org/articles/functional-green-tribunal-article-kanchi-kohli-mylawnet (Accessed: 5 January 2021).

Kukreti, I. (2019) "Environmental offences in India soared 790% in 2017," *Down to Earth*, 22 October 2019.

Malin, S.A., Ryder, S. and Lyra, M.G. (2019) "Environmental justice and natural resource extraction: Intersections of power, equity and access," *Environmental Sociology*, 5(2), 109–116.

Mathur, K. (2007) *Battling for Clean Environment: Supreme Court, Technocrats, and Populist Politics in Delhi*, CSLG/WP/03-01. New Delhi: Centre for the Study of Law and Governance, Jawaharlal Nehru University.

Motupalli, C. (2018) "Intergenerational justice, environmental law, and restorative justice," *Washington Journal of Environmental Law and Policy*, 8(2), 333–361.

Nambiar, S. (2012) "Paradigm of 'green' adjudication: Developing principles for Indian environmental decision-making in disputes involving scientific uncertainty," *ILI Law Review*, 1(1), 1–24.

OECD (Organisation for Economic Co-operation and Development) (2006) *Environmental Compliance and Enforcement in India: Rapid Assessment*. Paris. Available at: http://www.oecd.org/environment/outreach/37838061.pdf (Accessed: February 15, 2021).

Patel, K. and Dey, K. (2013) "The trajectory of environmental justice in India: Prospects and challenges for the National Green Tribunal" in Tim, M., Trivedi, N. and Vajpeyi, D. (eds.) *Perspectives on Governance and Society: Essays in Honour of Professor O.P. Dwivedi*. New Delhi: Rawat Publications, 160–174.

Patra, S.K. and Krishna, V.V. (2015) "National Green Tribunal and environmental justice in India," *Indian Journal of Geo-Marine Science*, 44(4), 445–453.

Peiris, G.L. (1991) "Public interest litigation in the Indian subcontinent: Current dimensions," *International and Comparative Law Quarterly*, 40(1), 66–90.

Pring, G. and Pring, C. (2009) *Greening Justice: Creating and Improving Environmental Courts and Tribunals*. New Delhi: The Access Initiative.

Pring, G. and Pring, C. (2016) *Environmental Courts and Tribunals: A Guide for Policy Makers*. Nairobi: UN Environment.

Radhakrishnan, R.K. (2015) "'Development' Disaster," *Frontline*, 25 December. Available at: http://www.frontline.in/cover-story/development-disaster/article7965568.ece?css=print (Accessed: 25 December 2015).

Rajamani, L. (2007) "Public interest environmental litigation in India: Exploring issues of access, participation, equity, effectiveness and sustainability," *Journal of Environmental Law*, 19(3), 293–321.

Raphael, C. (2019) *Engaged Scholarship for Environmental Justice: A Guide*. Santa Clara, CA: Santa Clara University.

Rengarajan, S., Palaniyappan, D., Ramachandran, P. and Ramachandran, R. (2018) "National Green Tribunal of India – An observation from environmental judgements," *Environmental Science and Pollution Research*, 25(12), 11313–11318.

Rosencranz, A. and Sahu, G. (2009) "National Green Tribunal bill, 2009: Proposals for improvement," *Economic and Political Weekly*, 44(48), 8–10.

Rosencranz, A., Sahu, G. and Raghuvanshi, V. (2009) "Whither the National Environment Appellate Authority?" *Economic and Political Weekly*, 44(35), 10–14.

Sahasranaman, P.B. (2012) *Handbook of Environmental Law*. New Delhi: Oxford University Press.

Sahu, G. (2008) "Implications of Indian Supreme Court's innovations for environmental jurisprudence," *Law, Environment and Development Journal*, 4(1), 1–19.

Sahu, G. (2014) "Why the underdogs came out ahead," *Economic and Political Weekly*, 49(4), 52–58.

Sahu, G. (2018) "Ecocide by design? Under Modi, vacancies at National Green Tribunal reach 70%," *The Wire*, 15 February. Available at: https://thewire.in/politics/ngt-political-apathy-vacancies (Accessed: 14 October 2020).

Schlosberg, D. (2004) "Reconceiving environmental justice: Global movements and political theories," *Environmental Politics*, 13(3), 517–540.

Schlosberg, D. and Carruthers, D. (2010) "Indigenous struggles, environmental justice, and community capabilities," *Global Environmental Politics*, 10(4), 12–35.

Seenivasan, R. (2014) *Law, Technology and Water Conflicts in Developing Societies: A Case Study of Tank Systems in Tamil Nadu*. PhD dissertation, University of Westminster, United Kingdom.

Seenivasan, R. (2016) "Judiciary and the destruction of Chennai's wetlands," *Economic and Political Weekly*, 51(50), 17–20.

Sharma, R. (2008) "Green courts in India: Strengthening environmental governance?" *Law, Environment, and Development Journal*, 4(1), 50–71.

Shekar, S. and Thomas, M. (2015) "Chennai's rain check: Fifteen years and counting," *Economic and Political Weekly*, 50(49), 19–21.

Suresh, M. and Narrain, S. (eds.) (2019) *The Shifting Scales of Justice: The Supreme Court in New Liberal India*. Hyderabad: Orient Blackswan.

Thompson, M. (2008) *Organizing and Disorganizing: A Dynamic and Non-Linear Theory of Institutional Emergence and Its Implications*. Devon: Triarchy Press.

Trubek, D.M. (1980) "Studying courts in context," *Law and Society Review*, 15(3/4), 485–501.

United Nations Department of Economic and Social Affairs (2020) *Sustainable Development Goal 16: Promote peaceful and inclusive societies for sustainable development, provide access to justice for all and build effective, accountable and inclusive institutions at all levels*. Available at: https://sdgs.un.org/goals/goal16 (Accessed: 29 August 2020).

United Nation Development Program (2020) *Global Multidimensional Poverty Index 2019: Illuminating Inequalities*. New York, NY: UNDP and OPHI. Available at: http://hdr.undp.org/sites/default/files/mpi_2019_publication.pdf (Accessed: 16 February 2021).

United Nations Economic and Social Commission for Asia and the Pacific (2008) *What Is Good Governance?* Bangkok, Thailand. Available at: http://www.unescap.org/pdd/prs/ProjectActivities/Ongoing/gg/governance.asp (Accessed: 29 August 2020).

United Nations Economic Commission for Europe (2019) *Convention on Access to Information, Public Participation in Decision-Making and Access to Justice in Environmental Matters*. United Nations Economic Commission for Europe. Available at: https://www.unece.org/env/pp/contentofaarhus.html (Accessed: 14 January 2019).

Véron, R. (2006) "Remaking urban environments: The political ecology of air pollution in Delhi," *Environment and Planning A: Economy and Space*, 38(11), 2093–2109.

Whyte, K. (2018) "The recognition paradigm of environmental injustice" in Holifield, R., Chakraborty, J. and Walker, G. (eds.) *The Routledge Handbook of Environmental Justice*. London: Routledge, 113–123.

Williams, G. and Mawdsley, E. (2006) "Postcolonial environmental justice: Government and governance in India," *Geoforum*, 37(5), 660–670.

World Bank (2020) *Poverty and Equity Data Portal: India*. World Bank. Available at: http://povertydata.worldbank.org/poverty/country/IND (Accessed: 8 October 2020).

World Commission on Environment and Development (1987) *Our Common Future*. Oxford: Oxford University Press.

Part II

Economic, policy and institutional context

2 Injustice and justice

The double movement of small hydro development in Himachal Pradesh

J. Mark Baker

Introduction

In the summer of 2012, protesting farmers in the northern Indian state of Himachal Pradesh (HP), in the Western Himalaya, forced a halt to construction work associated with a small, private, run-of-the-river hydropower project. The project would have disrupted the water flow into a nearby kuhl (farmer-managed irrigation system) that provides irrigation water to approximately 2,000 households in more than eight different villages. Letters outlining farmers' concerns sent to the hydropower company, the Deputy Commissioner, and even to the Chief Minister by the panchayat pradhan and Kuhl Committee President had produced no results, so local community members took matters into their own hands – they blocked further construction with threats to disable the partially completed infrastructure and engage in civil disobedience. Negotiations, which were ultimately successful, ensued. In early 2016, the Kuhl Committee President wrote to this author that "[w]e have successfully fought the case. The government and hydel project company conceded to all our conditions. Consequently, our society emerged 100% victorious."

This is just one example of the numerous instances of local opposition to the dozens of private run-of-the-river power projects across HP that are under construction or in operation. Opposition includes non-violent direct action, petitioning elected representatives and civil administrators, mobilizing support by developing coalitions with other communities and activist organizations and seeking redress through the judicial system. Without exception, the opponents object to the projects' negative effects on local livelihoods and the environment, and they seek redress or mitigation for the social and environmental injustices associated with privatized small hydropower development.

This chapter begins with a brief overview of small hydropower generation. I then present several key concepts from critical perspectives on neoliberalism to frame the analysis of the environmental injustices – the actual socioecological impacts – of the 49 commissioned small hydropower projects that operated in HP in 2013 (Government of Himachal Pradesh, 2012a,b), when this research was conducted. I then present two instances of collective

DOI: 10.4324/9781003141228-2

governance of small hydropower generation that might represent more justice-oriented approaches.

Analysing the neoliberalization of small hydro generation

Small hydropower projects utilize run-of-the-river power generation technologies to convert hydropower into electricity. These projects divert water from a source stream or river through a trench weir. After passing through a desilting tank, a power channel conveys the water along a contour line for approximately one to eight kilometres. The water then flows into a forebay and down a steeply sloped penstock to the powerhouse, where it drives one or more turbines. A tailrace returns the water to the source stream, from which a trench weir, further downstream, re-diverts the water for another power project. A complex set of operating controls monitors and manages the electricity the turbines produce. Power lines 1–15 km in length convey the generated power to the nearest HP State Electricity Board substation, at which point the power joins the state's power grid. Private corporations build, operate and maintain these hydropower projects. They sell the project's electricity to the HP State Electricity Board at guaranteed rates.

Because of the presumed local, regional and global socioeconomic and environmental benefits of small hydro, in the late 1990s, HP launched a series of initiatives to privatize and promote small hydropower production (Sinclair, 2003). The state's renewable energy agency, Himurja, was charged with overseeing the private development of the state's small hydropower potential. The state formalized the processes and mechanisms that govern private-sector involvement in electricity production by passing the Hydropower Policy 2006 (Government of Himachal Pradesh, 2006). By 2020, there were 88 commissioned projects, another 31 were under construction and approximately 620 were in various stages of planning (Chapter 7, this volume).

Key concepts from critical social science scholarship on neoliberalism help to frame and analyse the socioecological effects of privatized small hydropower in HP. These concepts concern state rollback and re-regulation, privatization and accumulation by dispossession, Polanyi's concept of double movement and proletarianization. I utilize these concepts within a notion of the state that acknowledges its internal heterogeneity, contingency and potential for supporting initiatives at odds with neoliberalism. This opens up an analytical space to consider the two exceptions to neoliberalism discussed near the end of this chapter.

The concepts of state rollback and re-regulation focus attention on the diverse and seemingly contradictory roles of the state within neoliberal environments. A common theme of neoliberal governance is deregulation or rollback of the state's functions in areas of cultural, social, economic or environmental significance. As with the logic behind privatization, neoliberal discourse holds that inefficient government bureaucracy, red tape and regulatory apparatus impede the realization of the full potential of individual creativity and so should be limited or reduced. Reducing the role of the state in a particular arena of life opens up the possibility of creating quasi-state or state-sanctioned private entities that perform functions that the state previously performed (Castree, 2010).

Closely related to state rollback is the notion of re-regulation, which refers to market-friendly state actions that support privatization and marketization. Markets do not operate in a vacuum, but rather they depend on particular forms of supportive state intervention. The centrality of state involvement in primitive accumulation runs counter to conventional wisdom understandings of neoliberalism as market rollback. As Erik Swyngedouw notes, "one of the central myths of the neoliberal model [is] that privatization means getting the state off the back of the economy and rolling back regulatory red tape" (2007, 55–56). In a similar vein, Nancy Peluso argues that the enclosure of commonly managed resources associated with marketization often entails public-private or state-capital alliances (2007, 89). The particular nature of this alliance provides insight into the specific roles the state plays in neoliberalizing projects and on the importance of identifying those roles.

The second set of concepts includes privatization, one of the hallmark characteristics of neoliberal approaches. Generally legitimized through appeals to efficiency, it entails assigning clear and enforceable private property rights to unowned, state-owned or communally owned elements of the social, cultural and natural world (Castree, 2010). Closely associated with privatization is marketization, which is the process whereby elements of the world that were previously not subject to a market logic become alienable and exchangeable.

Together, privatization and marketization can function as a form of what Harvey (2003) calls accumulation by dispossession. Accumulation by dispossession draws on the Marxist concept of primitive accumulation – the process of divorcing the producer from the means of production, changing property relations, consolidating capital and transforming human-environment relations to conform to a capitalist mode of economic production. Accumulation by dispossession refers to the ways in which profit accumulation by capitalist entrepreneurs is generated through dispossessing others, particularly communities, of resources that were often held in common and not part of capitalist circulations of monetized value and profit.

The third set of ideas revolves around Karl Polanyi's concept of double movement (1944). Polanyi argued that the expansion of market relations into new arenas simultaneously creates the conditions for the emergence of social movements that seek to counter the negative effects of market expansion. The double movement, thus, refers to the co-dependent processes of market expansion and community mobilization in opposition to such expansion. Market expansion can refer to the penetration of new geographic spaces by capitalist relations of production or to the expansion of such relations of production into previously non-capitalist arenas of production in the same space.

The fourth key concept concerns proletarianization, the process by which people are incorporated as workers into capitalist relations of production when such relations expand into new geographic spaces and sectors. Proletarianization is a central product of accumulation. When people are deprived of or separated from the means of production, they turn to wage labour. Proletarianization is a complex process, with workers at times resisting full proletarianization and at times struggling to gain access to more complete proletarianization. Within the context of small hydropower development, the concept of proletarianization draws attention to the labour relations within new private enterprises.

State rollback and re-regulation: No Objection Certificates(NOC) and the local area development authority

Small hydropower generation in HP relies on well-articulated relations between private power producers and the HP government. While the state may have receded in terms of its role as the sole power producer, a whole new set of regulatory approaches have emerged. The resulting dense network of highly structured relationships between the state government and independent power producers illustrates well the ideas of re-regulation that scholars of neoliberalism argue accompanies state rollback. As the discussion below demonstrates, re-regulation favours private power developers and state interests over rural communities.

The primary state entities that regulate small hydropower are the district civil administration, the HP State Electricity Board and Himurja. Himurja regulates and governs the process of private power producers developing small hydro sites. Small hydropower projects are exempt from the formal environmental assessment and public participation process required of larger projects (Diduck and Sinclair, 2016). Himurja does require developers to submit a variety of detailed reports concerning the economic, hydrologic, geologic and technical aspects. Additionally, at least for the first several years following approval of the Hydropower Policy 2006, the developer was also required to obtain No Objection Certificates (NOCs) from the government departments and village panchayats, the jurisdiction or interests of which the proposed power project would affect. These entities could include the Forest Department, Revenue Department, Irrigation and Public Health Department, Fisheries Department, Public Works Department, Pollution Control Board and Village Panchayat Councils. The purpose of the NOC was to identify the anticipated effects of the proposed project and develop mitigation strategies acceptable to the entity granting the certificate.

However, some government agencies grew reluctant to use the NOC process to mitigate negative anticipated effects. For example, one Fisheries Department officer told me that "before, we restricted small hydro projects, we tried to not give them NOCs, but we had to… it is the policy of the government" (to support hydropower development). Departments were under pressure to approve NOCs and/or request little or no mitigation. Similarly, Panchayat Councils that sought promises of remediation measures for anticipated negative consequences before granting an NOC or that attempted to withhold an NOC altogether came under intense pressure (Asher, 2008). Developers resorted to making informal payments to obtain the required NOCs (approximately 50 lakhs per MW to obtain the needed NOCs from each government department and 10–15 lakhs to obtain the NOC from a village pradhan) (Baker, 2014). Even with its shortcomings, the NOC process did provide an important mechanism by which government departments and local panchayats could require developers to address their concerns. Unfortunately, in 2014, the HP government eliminated NOCs from the formal approval process. The elimination of NOCs accompanied other changes aimed at streamlining the approval process, simultaneously making it more difficult for local communities and state departments to oppose small hydropower projects or require mitigation (Thakur and Asher, 2015).

The local district civil administration is also involved with the re-regulation of privatized small hydropower production. The Hydropower Policy 2006 requires project developers to deposit 1 per cent of the project cost into an account with the district commissioner (Government of Himachal Pradesh, 2006). These funds, known as Local Area Development Funds, are to be allocated by the local area development authority (LADA) to support local development activities, particularly those related to infrastructure and services.

The LADA programme was not working as well as intended. Most district revenue department offices in the state did not keep a centralized record of LADA obligations or project developer payments. This made it extremely difficult to hold project developers accountable for their LADA payment obligations. Consequently, many project developers did not contribute the legally mandated percentage, and district administrators seemed to have little authority or recourse to compel the project developer to do so. A related concern was the unevenness of awareness about the LADA programme among village pradhans. Several pradhans, especially in the remote areas of the state, had never heard of the LADA programme, even though one or more small hydropower projects were located within their panchayat boundaries.

Privatization and primitive accumulation

The development of small hydropower privatizes previously non-commodified community-owned and managed resources, which in many cases helped sustain rural livelihoods. This section explores the ways in which small hydropower dispossesses rural farmers, watermill operators, trout farmers and graziers. Many of these examples of dispossession are gendered.

Kangra District: disruption to local irrigation systems

Most of Kangra District lies on the southern side of the Dhaula Dhar mountain range, from where it extends across Kangra Valley and into the Sivalik Hills (Figure 2.1). The district is notable for its extensive network of community-managed gravity flow irrigation systems (kuhls). In Kangra Valley alone, 750 large and more than 2,500 small kuhls irrigate approximately 30,000 ha (Baker, 2005). Kuhl irrigation water is crucial for both kharif crops (rice and corn) and rabi crops (wheat and potatoes). These crops, except for potatoes, are almost entirely used for home consumption. Kuhl irrigation water is also essential for driving water-powered mills (gharats) and other machines, as well as irrigating home gardens, watering livestock and meeting household needs for non-potable water.

The importance of ensuring the continuity of these kuhl irrigation systems is reflected in the language of the NOCs that project developers had at one time to obtain from the Irrigation and Public Health Department as well as from village panchayat pradhans. These certificates contained language intended to protect community-managed kuhls from disruptions by small

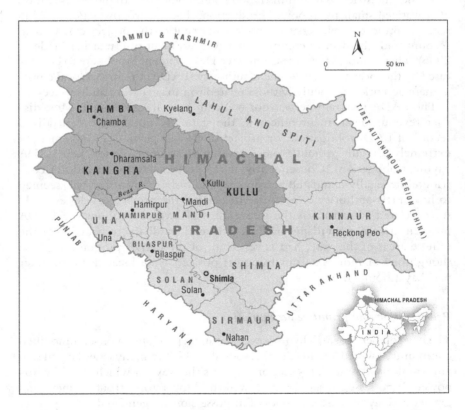

Figure 2.1 Map of Himachal Pradesh, highlighting Chamba, Kangra and Kullu districts.

hydropower projects and required the developer to pay full compensation if a project damaged or disrupted a community-managed kuhl.

Despite the protections delineated in the NOCs, small hydropower projects commonly disrupted kuhl irrigation systems or caused them to cease functioning altogether, either by physically damaging the irrigation system or by diverting the water on which the irrigation system relies. Small hydropower projects have disabled a total of 13 kuhl irrigation systems; in none of these cases did the project developer compensate farmers for their losses. This level of disturbance to irrigation is significant – for example, one of the disabled kuhls was the primary source of irrigation water for approximately 2,000 households (Baker, 2014).

Chamba District: landslides and damaged watermills

Chamba District lies to the north of the Kangra District and contains the headwaters of the Ravi River and key tributaries, all of which have cut deeply into the Himalayan Mountains. Because it lacks the broad arable plains that characterize the kuhl-irrigated Kangra Valley, farmers in Chamba combine rainfed cultivation on terraced fields carved into steep slopes with a high level of dependence on timber and non-timber forest resources, which meet both subsistence needs and generate revenue. The streams that flow from the forests down through the cultivated fields and villages to eventually join the Ravi River often power 10, 20, or more gharats (water-powered mills).

In steep, geologically unstable terrain such as this, small hydropower projects undermine local livelihoods by triggering landslides, damaging grazing land and cultivated areas and disabling or destroying local watermills. For example, landslides destroyed key portions of one small hydropower project on the Terailla stream course after it was commissioned in 2007, even though the developer's report stated the site was geologically stable. The upper edge of the growing landslide continues to move upslope and has harmed the common grazing grounds of the adjacent village.

Damage to gharats from small hydropower projects also commonly occurs in Chamba. Gharats are the most common method for grinding corn, wheat and occasionally rice. In exchange for grinding neighbours' grain, the gharat owner usually receives 10 per cent of the volume of grain they grind. These in-kind payments support the gharat owner's family; it is not uncommon for a woman to manage a gharat, often a widow or a woman head of household.

As of 2013, commissioned small hydropower projects had disabled a total of 104 gharats, either by destroying them due to land and rockslides or by diverting so much water that the gharat had to be abandoned due to lack of water (Baker, 2014). The elimination of these gharats weakened the economic stability of the large number of households whose livelihoods the gharats previously sustained. Although the Irrigation and Public Health Department NOC directed the power developer to provide adequate water flows for gharats, the Hydropower Policy contains no requirement that compensation be paid to gharat owners if the project damages their gharat or restricts the water available for diversion (Government of Himachal Pradesh, 2006).

Kullu District: interlinked aesthetic and economic effects

Kullu District's fame, which extends throughout India and indeed the world, stems from a variety of characteristics that also influence the pattern of socio-economic and environmental consequences of small hydropower development. These characteristics include picturesque mountainous scenery, favourable conditions for horticulture, important pilgrimage sites, fisheries development opportunities and regions with high conservation values. The diverse elements of the economic foundations of the district – fruit cultivation, commercial film production, tourism, pilgrimage, fisheries opportunities (stocking and trout farming) and conservation values – also heighten the stakes associated with the proliferation of hydropower projects. The commissioned small hydropower projects in the district accumulate private profit while dispossessing residents of common aesthetic and conservation values of the region.

Many Kullu residents have linked the area's natural beauty with the tourism and film industry and are worried about the negative effects on it of hydropower development. Residents decried the ugly transmission lines that cut through the fruit orchards in the main Kullu Valley and also traverse the deodar forests and cultivated areas of the tributary watersheds of the Beas River. One local film production manager noted ruefully that the density of transmission lines in the valley has already disrupted shooting operations and challenges the ability of film crews to obtain sequences not marred by transmission lines. Seeing the damage to apple orchards from transmission line construction and the fact that at least one person has died from electrocution from a low-hanging power line, families that own land where towers need to be constructed are increasingly reluctant to sell the small plot of land necessary to construct the power tower.

The negative effects of small hydropower development on water quality are also particularly evident in Kullu District. Project managers clean desilting tanks by flushing the accumulated silt directly back into the source stream, thus, creating a slug of sediment that harms downstream water quality and aquatic habitat and turns tap water the colour of mud.

These sediment slugs negatively affect downstream fisheries operations, both private and government. The Himachal Pradesh Fisheries Department's oldest trout hatchery is located at Patlikuhl in Kullu Valley. In 2009–2010, it produced 3.75 lakhs of fish ova, 80 metric tons of fish feed (sold to local fish farmers and as far away as Sikkim, Bhutan and Uttarakhand) and 12 metric tons of fish (Fisheries Department records 2012). This fish hatchery operation anchors the state's fish stocking programme and supplies fingerlings and other inputs to the growing number of households in Kullu that have established fish farming operations. The hatchery depends on clean, cold, oxygenated water to successfully manage the large number of tanks where fish eggs are fertilized and the ova are reared to become fingerlings or adults. Already, commissioned power projects have impaired the water quality of the hatchery, and more projects are planned. Hatchery managers are concerned about the threats to their source water posed by upstream hydropower development; they have written letters expressing this concern to the Director of the Fisheries Department.

Private power development dispossesses rural communities of the resources that sustain their livelihoods. The profit that accrues to producers from power generation is predicated on this dispossession, whether it be the dewatering of community-based irrigation systems, disabling of gharats that depend on common water, disturbance to communal grazing areas, undermining of fisheries-based livelihoods or impairment of important regional attributes of environmental aesthetics. Accumulation by dispossession is a primary mechanism of the environmental injustice of small hydropower. In response, rural communities in HP have assiduously resisted these forms of environmental injustice.

Resistance and community opposition – The double movement

Almost every example of privatization and dispossession described above has been met with a counter-effort to either obtain redress or compensation or to halt the project entirely. Rural people seek justice through a diverse array of mechanisms and strategies. These include petitioning civil administrative authorities and the power producers themselves, direct action and seeking redress in the judicial arena, including the National Green Tribunal. All of these strategies exemplify what Polanyi called the double movement: the collective efforts of communities to push back against the social, economic and environmental degradation caused by privatization, marketization and the expansion of capitalist relations of production into spaces and sectors where it had previously not existed.

When small hydropower projects dewater kuhl irrigation systems, local farmers have not stood by idly. Farmers have made countless visits to district administrative authorities petitioning them to intercede on their behalf to seek redress, compensation and/or release of adequate water flows necessary for irrigation (Baker, 2014). Despite these requests, there has not been one instance in which the district administration prevailed upon the power producer to either compensate for disruptions to these irrigation systems or reduce their water diversion to leave enough water for kuhl irrigation. While this chapter began with a successful example of farmer direct action in defence of kuhl irrigation water, in most instances, farmers whose kuhls are dewatered by small hydro projects receive no mitigation or compensation.

Community resistance to small hydropower development has been particularly forceful and effective in Chamba District. Local community members, mostly women, whose grazing lands have been threatened by landslides associated with the Terailla hydropower project have protested on numerous occasions and filed multiple court cases against this and three other adjacent projects. Some of these protesters, including local village women, have been arrested and detained overnight in jail.

Gharat owners in a nearby watershed, seeing the pattern of uncompensated damage when gharats were disabled due to landslides or dewatering, decided on a proactive strategy. For six months, using threats of direct action against a newly commissioned small hydropower project, the owners of 12 project-affected gharats stopped the power project from operating until an acceptable compensation agreement was successfully negotiated. After the first compensation payment had been made to the concerned gharat owners, they removed their opposition to the project and it began producing and selling electricity. However, as one gharat owner noted, if their payments cease, they will again stop the project through direct action.

The ability of these gharat owners to successfully engage in direct action and then negotiation reflects the pre-existing patterns of social activism and strong local governance traditions prevalent in Chamba. Local leaders, inspired by Gandhian ideologies of self-governance and sustainable local livelihoods, have worked to strengthen village panchayat institutions over the last two decades. This awareness building and social mobilization has centred on defending village community timber and non-timber forest product rights, advocating for community-based medicinal herb collection and strengthening village level democratic institutions (Gaul, 2001). The resulting awareness and knowledge concerning local rights and democratic process has empowered local communities to defend against livelihood threats, including threats from small hydropower projects.

This capacity to resist also informs the efforts of one village community to oppose a proposed small hydropower project in the area by submitting an application to the National Green Tribunal. The application names the state of HP and the project developer as respondents. It outlines various ways in which the current proliferation of small hydropower projects across the state disrupts local livelihoods and environmental features and, citing the impacts of two such projects in one watershed, it requests a halt to further construction based on the violation of legal frameworks such as the Environment (Protection) Act 1986, India's 2012 national water policy and Himachal Pradesh's own Hydropower Policy 2006.

In neighbouring Kullu District, community opposition focused on opposing the granting of NOCs and pursuing justice through the court system. The potential threat that small hydropower development poses to fish farming has strengthened local community opposition, which has manifested as local panchayat refusal to grant the NOC.

The most successful example of local opposition to small hydropower development is the resistance that emerged in the Tirthan Valley, part of which is in the Great Himalaya National Park. The valley is famous for its natural beauty as well as world-class sport fishing. Dilaram Shabab, a retired Member of the Legislative Assembly (MLA) from this area, spearheaded the opposition to small hydropower development. With the eventual support of local panchayats, community members and fishing lodge owners, as well as backing from Fisheries Department, Forest Department and Great Himalayan National Park officials, a five-year court battle was launched against small hydropower development in the watershed. After three years of arguments and rulings in the Kullu District Court and more than one year in the High Court in Shimla, the High Court presiding judge ruled in favour of the arguments set forth concerning the negative effects on the environment, fisheries and affected communities of the planned small hydropower projects in the watershed. The High Court declared the Tirthan off limits to all hydropower projects, and it cancelled the nine previously approved small hydropower projects (Civil Writ Petition 1038, 2006).

These diverse examples of resistance and opposition to the expansion of privatized small hydropower development are grounded in the shared defence of rural livelihoods and the environments that support those livelihoods. They defend livelihoods and lifeways that enable material, cultural and social reproduction. In some instances, livelihood struggles and workers' rights struggles emerge within the same context, as the following discussion of proletarianization illustrates.

Proletarianization and small hydropower

Because the lack of local employment opportunities is one of the primary drivers of migration from hill areas, the provision of permanent jobs through small hydropower projects could be a significant benefit. In addition to a steady income, permanent regular employees participate in government-approved pension plans, receive compensation for work-related accidents and injuries and are protected from arbitrary dismissal. Recognizing the importance of generating local employment, the Hydropower Policy stipulates that 70 per cent of the project's workers be from HP (Government of Himachal Pradesh, 2006). The project developer is also required to register all workers with the Labour Department and the local police station on a monthly basis.

The 49 commissioned small hydropower projects in the state at the time of this research did generate significant employment. However, the quality of that employment was often poor, with egregious examples of labour abuse. More than half of project developers had evaded complying with labour law. Only 22 project developers had registered their employees with the state Labour Department as regular employees. These workers received the protections and benefits of the state's labour laws. However, the workers in the remaining 27 small hydropower projects, while doing the same work as regular employees in other projects, were hired on a daily wage basis and were, thus, excluded from the benefits and security of regular employment. Only 11 project developers had established provident fund contributions for their employees; the remaining 38 had not. For the majority of workers in small hydropower projects, one of the most important potential local benefits – secure employment – was not realized.

Given the significant risk of injury or death in this sector, it is of particular concern that unregistered workers are less likely than registered workers to receive compensation should an accident occur (Himdhara, Environment Research and Action Collective, 2019). Tragically, 40 people have died in accidents related to the construction of the 49 commissioned small hydropower projects examined in this study. Most of these deaths occurred during the project construction phase due to tunnel collapses, falling rock, landslides and tractor accidents. Only three of the families of the 40 workers who died in fatal accidents received some form of compensation. The lack of proper registration and the general absence of compensation suggest the extent to which many of the projects treat workers as a disposable labour force.

The common practice of contracting out project construction work to subcontractors, who hire large numbers of employees, also challenges the ability of unions to advocate for project workers. Questions arise concerning who is ultimately responsible for following the relevant labour laws and protections – the project developer or the developer's subcontractors? Project developers evade accountability through the use of subcontractors or by creating subsidiary companies. These strategies challenge the ability of unions to seek redress for labour law violations and demand worker rights.

The uneven or partial proletarianization of the workforce in small hydropower projects reflects the contradictory processes of accumulation by dispossession and capitalist expansion. Rural residents depend on and value wage labour to supplement subsistence forms of livelihood reproduction. Yet simultaneously, subsistence forms of production subsidize household reproduction. This allows employers to keep wages low and extract more surplus value, despite labour laws that specify the higher wages and social benefits that should accrue when workers perform the tasks associated with regular rather than casual employment.

Two exceptions to neoliberalism

Two examples of collective governance of small hydropower development stand in stark contrast to the general patterns discussed above. The Sai Engineering Foundation and the Churah Floriculture Cooperative Society both represent exceptions to neoliberalism because, though they operate within a neoliberalized sector, they function according to different logics and underlying purposes than profit. The work of these two entities demonstrates the importance of being attentive to the fractured and potentially contradictory nature of the neoliberal state (Mansfield, 2004).

Inspired by the teachings of the religious leader Bhagwan Sri Sathya Sai Baba and the religious ideals of Gandhian social service, Sai Engineering Foundation is a registered charitable foundation that promotes social welfare. The foundation has been involved with hydropower development since the first India Hilly Hydel demonstration projects. The Sai Foundation owns and manages its own projects and provides consulting services for other private

power developers. The foundation invests revenue from hydropower production in social service and welfare programmes in HP (Sai Engineering
Foundation, 2011). Because of the social service ideology that informs this
organization, when the Sai Foundation develops small hydropower projects,
it does so in a manner that does not generate the negative socioecological
consequences associated with other project developers.

The second exception to neoliberalism is the Churah Floriculture
Cooperative Society. The Hydropower Policy 2006 specifically addresses
the need to prioritize working with cooperative societies. This reflects the
government's historic association with more socialist models of development
and the long-standing traditions of formally registered cooperative societies, which date back to the 1860 Societies Registration Act. Since 1996, the
Churah Cooperative Society has worked to promote the economic development of low-income families in the Churah Valley, a remote area in Chamba
District, not far from the border with Jammu and Kashmir. The cooperative's
initial and ongoing work involves developing floriculture using greenhouses
and marketing cut flowers to cities in north India, as well as producing
off-season vegetable crops in neighbouring Pangi Valley. The cooperative is
also working to develop a small hydropower project under the framework of the
Hydropower Policy 2006. Himurja allotted the cooperative society a 5-MW
small hydro project in 2009. By 2014, all necessary studies were completed,
the NOCs and other clearances had been obtained and the Implementation
Agreement with Himurja had been signed. Interestingly, 400 Below Poverty
Line households, all members of the cooperative society, are involved in this
effort. In order to qualify for the necessary loans, each household put up its
house and land as collateral. The revenue from the small hydropower project,
once it is commissioned, will be shared among the participating families.

Both the Sai Engineering Foundation and the Churah Floriculture
Cooperative Society represent viable alternatives to corporate ownership of
small hydropower facilities. Both of these organizations are accountable to
local concerns and interests and prioritize local social and environmental sustainability. However, both the foundation and the cooperative face an uphill
battle to get their projects approved. Both organizations have minimal financial resources and are, thus, at a disadvantage when competing with private
corporations for bureaucrats' attention and project approvals.

The Sai Engineering Foundation and the Churah Cooperative Society
represent important exceptions to neoliberalism that are more consistent
with notions of environmental justice. They demonstrate that the local man-
ifestations of neoliberalism are grounded in particular social, historical and
cultural contexts (Brenner and Theodore, 2002; Peck and Tickell, 2002).
These contexts can create spaces for alternative models or ways of organizing
that may challenge or resist dominant neoliberal ideology and policy. These
alternative models exist due to the contingent and somewhat fractured nature
of state authority. In this case, the state's historic commitments to cooperative
societies as a model of national development, the cultural idioms of Gandhian

forms of social service and civic engagement and the religious notions of spiritual practice through right action combine to create institutional space for these two exceptions to neoliberalism.

Conclusion: seeking environmental justice in a neoliberal environment

Neoliberal approaches to small hydropower have negatively affected household reproduction and the environments on which rural communities depend, within both capitalist and non-capitalist modes of production. Resistance has emerged to defend rural lifeways and demand enforcement of labour laws and regulations. Struggles to secure workers' rights and pushback against neoliberalization's "parasitization of formally non-capitalist processes" (Glassman, 2006, 617) are at the heart of local efforts to work towards environmental justice. However, these multifaceted forms of resistance against the diverse harms of small hydropower development are not necessarily coordinated or integrated. So, for example, the efforts of labour unions to secure workers' rights by striking and the opposition of irrigators of the nearby kuhl to the adjacent power project were not coordinated, despite that fact these processes were occurring within the same small watershed and within contiguous villages, in opposition to the same phenomena. Similarly, the application submitted to the National Green Tribunal against a proposed small hydropower project and the state of HP focused only on the socioecological effects on rural community lifeways. It did not include an analysis of problems associated with the proletarianized workforce within neighbouring hydropower projects.

Farmers' struggles to secure reliable water supplies and labour union efforts to address labour law violations are examples of environmental justice struggles against the injustices of privatized small hydropower development. The effectiveness of each effort would be enhanced through joining in common cause. Doing so requires an analysis of the injustices of this form of power development that incorporates its effects within both capitalist and non-capitalist modes of production. Developing such an analysis would facilitate achieving environmental justice across multiple sectors and modes of production.

Lastly, collective governance of small hydropower projects with a focus on social welfare, illustrated by the Churah Floriculture Cooperative Society and the Sai Engineering Foundation, represent promising exceptions to the neoliberal model of privatized small hydropower development. Supportive state policy encouraging these and similar exceptions could potentially strengthen the environmental and social justice dimensions of small hydropower.

References

Asher, M. (2008) "Impacts of the Proposed Himalayan Ski-Village Project in Kullu, Himachal Pradesh – A Preliminary Fact Finding Report." Himachal Pradesh: Him Niti and Jan Jagran Evan Vikas Samiti.

Baker, J.M. (2005) *The Kuhls of Kangra: Community-Managed Irrigation in the Western Himalaya*. Seattle, WA: University of Washington Press.

Baker, J.M. (2014) "Small hydropower development in Himachal Pradesh: An analysis of socioecological effects," *Economic and Political Weekly*, 49(21), 77–86.

Brenner, N. and Theodore, N. (2002) "Cities and the geographies of 'actually existing neoliberalism,'" *Antipode*, 34(3), 349–379.

Castree, N. (2010) "Neoliberalism and the biophysical environment: A synthesis and evaluation of research," *Environment and Society: Advances in Research*, 1(1), 5–45.

Diduck, A. and Sinclair, J. (2016) "Small hydro development in the Indian Himalaya: Implications for environmental assessment reform," *Journal of Environmental Assessment, Policy and Management*, 18(2), 1–24.

Gaul, K. (2001) "On the move: Shifting strategies in environmental activism in Chamba District of Himachal Pradesh," *Himalaya*, 21(2), 70–78.

Glassman, J. (2006) "Primitive accumulation, accumulation by dispossession, accumulation by 'extra-economic' means," *Progress in Human Geography*, 30(5), 608–625.

Government of Himachal Pradesh (2006) *Hydro Power Policy*. Shimla: Government of Himachal Pradesh.

Government of Himachal Pradesh (2012a) Memorandums of Understanding, Himachal Pradesh Energy Development Agency (Himurja). Available at: http://himurja.nic.in/mountilldate.html (Accessed: 25 May 2012).

Government of Himachal Pradesh (2012b) Himachal Pradesh Energy Development Agency. Available at: http://himurja.nic.in (Accessed: 25 January 2013).

Harvey, D. (2003) *The New Imperialism: Accumulation by Dispossession*. Oxford: Oxford University Press.

Himdhara, Environment Research and Action Collective (2019) *The Hidden Costs of Hydropower*. Available at: http://www.himdhara.org/wp-content/uploads/2019/06/The-Hidden-Cost-of-Hydropower_2019.pdf (Accessed: 4 November 2020).

Mansfield, B. (2004) "Rules of privatization: Contradictions in neoliberal regulation of North pacific fisheries," *Annals of the Association of American Geographers*, 94(3), 565–584.

Peck, J. and Tickell, A. (2002) "Neoliberalizing space," *Antipode*, 34(3), 380–404.

Peluso, N. (2007) 'Enclosure and privatization of neoliberal environments' in Heynen, N., McCarthy, J., Prudham, S. and Robbins, P. (eds.) *Neoliberal Environments: False Promises and Unnatural Consequences*. New York, NY: Routledge, 89–93.

Polanyi, K. [1944] (1957) *The Great Transformation: The Political and Economic Origins of Our Time*. Boston, MA: Beacon Press.

Sai Engineering Foundation (2011) "Karmayoga," *Quarterly Newsletter of Sai Engineering Foundation*, 1(11). New Shimla, Himachal Pradesh: Sai Engineering Foundation.

Sinclair, J. (2003) "Assessing the impacts of micro-hydro development in Kullu District, Himachal Pradesh, India," *Mountain Research and Development*, 23(1), 11–13.

Swyngedouw, E. (2007) "Dispossessing H_2O: The contested terrain of water privatization" in Heynen, N., McCarthy, J., Prudham, S. and Robbins, P. (eds.) *Neoliberal Environments: False Promises and Unnatural Consequences*. New York, NY: Routledge, 51–62.

Thakur, K. and Asher, M. (2015) "A Himalayan sellout" (blog), *Down to Earth*, 30 November. Available at: https://www.downtoearth.org.in/blog/a-himalayan-sell-out-48780 (Accessed: 30 October 2020).

3 The visible fault line of development

The right of consent of Adivasi communities and the political economy of mining

Arpitha Kodiveri

Introduction

"We will not give our land away, it is the abode of Niyamraja, the lord of law," says Lado Sikaka, leader of the protest against the acquisition of his community's sacred land for bauxite mining by Vedanta Resources Limited (Bera, 2013). The government of Odisha, a mineral-rich state in eastern India, has justified this acquisition as a part of the development process. The pitting of interests of Adivasi (Indigenous) communities against the state and mineral corporations has been central to the development narrative of India's emerging economy. This conflict is a microcosm of the larger questions facing many emerging economies, encompassing issues such as reconciling development with democracy, environment and human rights standards with deregulation and fast-paced growth with rising inequality.

In this chapter, I investigate the dynamics that surround this and other conflicts over land rights in mining areas. The legal landscape in which this conflict is situated was reconfigured in 2013, when the Supreme Court in a case against Vedanta recognized the right of consent of Adivasi communities under the Forest Rights Act (FRA) 2006 (*Orissa Mining Corporation v Ministry of Environment and Forest and Others*, 2013). This was followed by an amendment to the colonial Land Acquisition Act 1894, in which the consent of landowners was made mandatory for acquisition. Prior to 2013, the consent of Adivasi communities was not recognized (Ramesh and Khan, 2015). A highly divisive issue, or fault line, is plaguing India's forests: forest-dwelling communities are unable to participate adequately in decisions about acquisition of forest land, resulting in forceful acquisition of these lands and resources.

Recognition of the right of prior, informed consent has made this fault line visible as communities have begun to assert their right under the FRA and the new land acquisition law (Ministry of Environment and Forests, 2009 circular; amended the Land Acquisition Act 2013).[1] The right of prior, informed consent is a legal requirement where, if the land is being acquired,

DOI: 10.4324/9781003141228-3

the Gram Sabha (village assembly) may collectively draft a resolution where they accept or reject such an acquisition. In the event of a rejection, the administrative authority, in this case the Forest Advisory Committee, a federal-level committee that decides on the diversion of forest land, has to take into account the rejection before granting land to the mining company. The land can be provided to a mining company despite having been rejected by the village assembly, but the land has to be granted through a process of negotiation, until the village assembly is convinced that their conditions for providing consent have been met. The question remains of how the decision made by the Gram Sabha impacts the terms and conditions of the state's eventual decision to provide land for mining. I argue that this decision is shaped by the state's predominant model of development, the political economy of mining, the nature of legal mobilization and the mining company involved in the conflict.

Here I trace the change in the legal landscape around legally recognizing and implementing the right of consent, through various phases in India's development story. I then locate the enforcement of this right in the political economy of mining in India. The central argument of this chapter is that implementing the right to prior, informed consent takes place in a context where state imperatives for mining are embedded in a pro-business model of development and an environment of increased deregulation.

In the following section, I trace India's development story and examine it in relation to changes in the democratic structure. This historical lens lays a foundation for the discussion in the third section on the regime of dispossession in India. The fourth section explores the nature of legal mobilization that has taken place in the Vedanta case, which involves a multinational enterprise based in London setting up an aluminium refinery in Lanjigarh, Odisha. The final section offers concluding comments.

Democracy, development and the state in India

Development and democracy from 1947 to the present

It is difficult to narrate the story of development in India without also tracing its path as a young democracy breaking out of the shackles of colonial rule. India's development can be viewed either through the lens of its accomplishments and high rate of economic growth or through the lens of increasing levels of inequality. I adopt the latter view to describe what many scholars have identified as economic growth made possible from accumulation by dispossession (Harvey, 2013). The dialectic of development and democracy is nested in the process of prior dispossession made possible by the colonial land and forest laws, which have now changed. Accumulation by dispossession serves as an effective starting point for this narrative.

The dialectic of democracy and development in India is central to understanding the country's political economy. A pertinent argument made in the literature on India's political economy relates to the changing nature of the state and its modality of decision-making in the different phases of development and democracy (Kohli, 2010). The nature of the state has shifted from a developmental orientation to one that becomes difficult to categorize, although it shares the characteristics of a developmental state with a neoliberal agenda. The changing nature of the state will enable us to frame the context within which the right of consent operates. At times, the developmental prerogative overtakes democratic ideals and in other instances, democratic ideals take precedence. The right to prior, informed consent is situated between the competing demands of democratic expression through participation in decision-making and the developmental prerogatives of the Indian state.

The narrative of development in India can be divided into four phases (Table 3.1), beginning with economic nationalism,[2] followed by the selective opening up of the Indian economy to foreign investment. The third phase is that of liberalization triggered by the structural adjustment program in 1991,[3] followed by the phase beginning in 2014 characterized by deregulation and the advent of the World Bank's ease of doing business reforms.[4] In each of these phases, the law has played a crucial role in enabling these development models to be realized. This division of the narrative of development and democracy is derived from Atul Kohli's analysis in his seminal work on development and democracy in India (Kohli, 2010).

Modi-nomics and ease of doing business

The present (fourth) phase of development, characterized by so-called Modi-nomics under Prime Minister Narendra Modi, is a continuation of increased liberalization with further opening of the economy. Sectors that were previously nationalized are now being opened to foreign direct investment (Panda, 2016). For example, in February 2018, an announcement was made to privatize coal and encourage foreign investors to participate in that sector (Cabinet Committee on Economic Affairs, 2018).

The increased liberalization that accompanies this change amounts to a new form of economic nationalism. This new economic nationalism features foreign direct investment allied with Indian business actors through partnerships (Panda, 2016). "Make in India" is a set of economic reforms proposed by the Modi government to boost local manufacturing and industrialization (Misra and Ghadai, 2016). These reforms are being done in consultation with the World Bank to improve India's ranking in the Ease of Doing Business Index. A closer examination of the reforms reveals an attempt to deregulate labour, environmental and social standards (Bajpaee, 2016).

Although the country's democratic structure remains the same, the Planning Commission, which directed India's development agenda, was abolished in 2014. The reason for its abolition was to increase cooperative federalism and increase the role of subnational governments in economic decision-making. In its place, the National Institute for Transforming India

Table 3.1 The four phases of development in India and associated time periods and changes in the democratic structure

Time period	Development model and policy	Democratic structure	Nature of the state
1947–1980	• Economic nationalism with policy of import substitution guiding industrialization • Formation of Planning Commission, which produces five-year plans to guide all aspects of economic decisions (Sinha, 2011)	• Foundation for democracy laid with the drafting of the Indian Constitution • Federal structure, with concentration of power respecting economic decisions at the federal level and not at the subnational level [5]	• Developmental state, with guiding the pace and direction of development processes (Caldentey, 2008)
1980–1991	• Foundations of pro–business development model • Highly interventionist state that begins to prioritize economic growth in the national interest (Kohli, 2010)	• Increased influence of business lobby in state decision-making; increased acquisition of land for large-scale development projects	• Developmental state, with a strengthened alliance with the business elite
1991–2014	• Beginning of liberalization of Indian economy, triggered by a structural adjustment programme	• Radical decentralization with the 73rd constitutional amendment [6] • Formation of local governments at the village level (panchayats)	• State's relationship with business elite drives economic decision-making (Kohli, 2010)
2014–	• Further liberalization of Indian economy, with radical economic reforms, including privatization of sectors that had remained nationalized, like coal • Dissolution of Planning Commission and formation of NITI Aayog • Resurgence of economic nationalism (Kohli and Menon, 2016), with new "Make in India" reforms	• Democratic structure remains the same	• State's relationship with business elite drives economic decision-making • NITI Aayog becomes nodal authority where these decisions are made (Cabinet Secretariat Resolution No. 511/2/1/2015)

(NITI Aayog) was established, with the agenda to act as a think tank that would provide guidance to the central and subnational governments on their development goals (Cabinet Secretariat Resolution No. 511/2/1/2015).

In this phase, the alliance between the state and business elite has been strengthened by the model of partnership used for foreign direct investment, a mechanism for the flow of global capital, particularly in the extractive sector. Another important phenomenon under this new brand of economic nationalism is increased scrutiny and criticism of civil society, in which opposition to developmental objectives has led to activists and others being labelled as anti-national elements. Furthermore, the legal advancements regarding the right of consent have been questioned on the basis that the right stands in the way of national developmental objectives (Make in India – Foreign Direct Investment, no date).

The pro-business development model persists with the objective of shifting the role of the state from that of a "player" in the industrial and service sectors to an "enabler" through the use of legislation, policymaking and regulation to support business activities (Cabinet Secretariat Resolution No. 511/2/1/2015). This is captured in the NITI Aayog founding document as "minimum government and maximum governance." Though this signals the state's retreat from direct provision of economic services and goods, it solidifies the state's role in shaping the future developmental narrative as an enabler. The developmental state now exists in a different form, where the state will use its legislative, executive and regulatory functions to enable growth and businesses while not partaking in business activities itself (Cabinet Secretariat Resolution No. 511/2/1/2015).

Regimes of dispossession in India

An essential element of India's development story is the changing land laws and processes of dispossession. Analysing it through this lens enables us to witness a shift in the legal regime from consultation of landowners to consent. In this section, I trace the regime of dispossession alongside the different phases of development, to compare the development models and their enabling legal frameworks.

Michael Levien defines a regime of dispossession as

> an institutionalized way of expropriating land from [its] current owners or users. It has two essential components: a state willing to dispossess for a particular set of economic purposes that are tied to particular class interests; and a way of generating compliance to this dispossession. These two aspects are intrinsically linked.
>
> (Levien, 2015, 149–150)

The willingness of the state to dispossess is a negotiation between its developmental and democratic interests. These aspects have changed over the

Table 3.2 The law and economic reasons in relation to corresponding phases of development

Phase of development	Regime of dispossession
Economic nationalism	• Doctrine of Eminent Domain used for acquisition of land as a process of nation building; this occurs when land reforms have not succeeded • Adivasi lands protected within the category of "scheduled areas"[7]
Foundations of a pro-business development model	• Use of "public purpose" clause under the Land Acquisition Act, 1894 for large-scale industrialization[8]
Liberalization	• Introduction of consultation with local communities as a part of the environmental impact assessment process[9] • Transition from mere consultation to requirement of consent with the new Land Acquisition Act, 2013 and the Forest Rights Act, 2006[10] • Land acquisition continues despite these safeguards against large-scale industrialization
Ease of doing business	• Dilution of the requirement of consent through the land ordinance (Kohli and Gupta, 2016)

different phases of development and the associated laws that enable such dispossession. Table 3.2 locates the law and set of economic reasons corresponding to each phase of development. The changes in the laws show a gradual shift from a process of seeking consultation from communities to consent and its dilution. The concept of regimes of dispossession invites us to see dispossession as a form of coercive redistribution that serves different purposes and class interests in different periods, but the contribution of which to economic change and the "public interest" is not only open to question but also subject to political contestation (Levien, 2015).

Ease of doing business and the regime of dispossession

The first legislative change that took place in relation to land laws in the ease of doing business phase was the removal of the consent provision under the Land Acquisition Act 2013. This was done through an ordinance. What has come to be known as the Land Ordinance was vehemently opposed by farmers, Adivasis and lower caste communities who are primarily dependent on land for their livelihood. This contentious ordinance was eventually not passed, though at the subnational level changes were made to the applicable land laws. Kanchi Kohli has addressed these changes where the consent provision has been removed in the process of land acquisition in seven subnational units (Kohli and Gupta, 2016).

Although this phase sees the state moving away from being an active player in economic activities, this shift does not apply to land acquisition. In a document titled "Land Acquisition: The Rationale and the Way Forward," Arvind Panagariya, Vice Chairman of NITI Aayog, suggests that the state should continue to act as the land broker, because since land ownership rights in India are poorly defined, companies will otherwise not be able to access large parcels of land as smaller disputed parcels will hamper their endeavours (Panagariya, 2015). This uncertainty would be overcome if the state were to play an active role in land acquisition where the project serves a sufficiently important public purpose. The determination of a project's importance would be left to the discretion of the state. The assertion of consent, according to Panagariya (2015, 76), is a "hold-up" problem that state intervention can overcome.

The consent requirement under the FRA 2006 was diluted through a notification whereby an exception to the consent requirement was made for linear projects like roads, highways and pipelines (Jebaraj, 2013). The requirement of consent for mining projects came to attention when in the state of Chhattisgarh, the district administration cancelled forest rights that were granted to Adivasi communities to make way for coal mining. It is in this contentious atmosphere that legal mobilization takes place, including the possibility of calling upon international law if a multinational company or international investment is involved (Kodiveri, 2017).

Legal mobilization

Mobilizing around rights has become the predominant means for social movements to make legal claims. This phenomenon is referred to as "rightful resistance" by scholars like Santos and Nandini Sundar (Santos and Rodríguez-Garavito, 2005; Sundar, 2009). Mobilization around rights and legal claims is an integral part of social mobilization strategies in India, and social movements have been successful in translating their legal asks into legislation such as the FRA (Jebaraj, 2013).

The use of law as an essential aspect of social mobilization in cases of dispossession has its limitations, as legal battles only sometimes result in a win. As Dash (2016) has pointed out, the state's selective interpretation of rights–based legislation means that in some cases rights acquire actual force but in other cases they are superseded by the primacy of the notion of eminent domain, making legal struggles a slippery territory. This situation is further compounded when states introduce newer laws or amendments to existing laws that dilute the pro-rights tone of such legislation. Given this chequered terrain and the political economy on which legal mobilization takes place, social movements rely on creative legal strategies and fora to challenge dispossession.

One such example is the strategies of legal mobilization adopted by the anti-Vedanta social movements against mining. Their strategies highlight the interaction between soft law (i.e., non-binding policies) and statutory law, and the opportunity that the involvement of multinational enterprises provides for social movements to bypass the state. This agility offers a route for inter-action between domestic law and international legal standards. Though India is a signatory to the United Nations Declaration on the Rights of Indigenous People (UNDRIP), it does not recognize the term "Indigenous." This lack of recognition has prevented Adivasi communities from mobilizing interna-tional law like UNDRIP.

Free, prior and informed consent and legal pluralism

Laws and standards relating to free, prior and informed consent (FPIC) are found in a myriad of sources, and FPIC has taken on its own language and rules. As Cesar Rodríguez-Garavito observes, the principle of FPIC

> has precipitated a true explosion of hard and soft law norms at both the international and national levels, which incorporate different versions of indigenous peoples' right to FPIC. In fact, I argue that FPIC's rise and impact in regulations and disputes about indigenous rights have been so profound that instead of merely constituting a legal figure, it entails a new approach to ethnic rights and multiculturalism, with its own lan-guage and rules.
>
> (Rodríguez-Garavito, 2010, 6)

The multiplicity of laws and standards coexists with overlapping notions of what constitutes FPIC. In different regulatory settings, FPIC carries different meanings and interpretations.[11] There are varied procedural safeguards for its realization, and it is this maze of legal pluralism that local communities mobi-lize under the broad umbrella of FPIC. The concept of legal pluralism is cap-tured aptly by Tamanaha (2007) when he states that legal pluralism recognizes coexisting or overlapping bodies of law and the diversity within them. Each of these bodies of law makes competing claims of authority and has different styles or orientations. This, when applied to FPIC, explains the complexity of regulatory structures and law engaged in its protection and recognition.

The precipitation and plurality of norms, laws and procedures are evident in the Indian context. The requirement of consent exists in domestic hard law, international law and soft law. The soft law norms also exist in the cor-porate social responsibility codes of mining and other companies (Berman, 2012). Table 3.3 shows the variety of laws and policies – the regulatory pluriverse – engaged in the governance of FPIC in the Vedanta case, which I examine more closely in the next section.

The regulatory pluriverse of FPIC and strategies of legal mobilization in the case of Vedanta Resources

The multinational Vedanta Resources Limited, through its wholly owned Indian subsidiary Sterlite Industries, entered into a memorandum of

Table 3.3 Laws and policies engaged in the governance of FPIC in the Vedanta case

Type of law or policy	Relevant FPIC governance in Vedanta case
Domestic law on FPIC, including associated case law	• FRA • Schedule V • Samatha judgement
International law	• UNDRIP
Soft law	• OECD guidelines for multinational companies • Ethical investment policies • Sustainable Development Framework (SDF)
Company codes of conduct or CSR policies	• Technical guidance documents on FPIC • Stakeholder engagement policies on FPIC

understanding with the state of Odisha for the establishment of an aluminium refinery in Lanjigarh in 2003. The bauxite for the aluminium refinery was to be mined from the nearby bauxite reserve in the Niyamgiri hills, and the state-owned Odisha Mining Corporation guaranteed the supply of bauxite from Niyamgiri. This agreement was met with opposition by the local forest-dwelling community on the grounds that their sacred rights to the hill would be violated. The strategy of these communities was to mobilize primarily against Vedanta Resources.

At the time when the memorandum of understanding was signed, there was no requirement for obtaining consent apart from the Samatha judgement. In this landmark 1997 judgement, the Supreme Court had ruled that land in scheduled areas cannot be sold to non-tribal community members, particularly to private mining companies (Samatha v State of Andhra Pradesh and Others Appeal, 1997). The guidelines set out in the Samatha judgement ensure that the land in scheduled areas cannot be alienated for private profits. As Niyamgiri was part of the scheduled area, the guidelines under this precedent were applicable. The mine was to be located in the forest area and, therefore, also required a forest clearance from the Ministry of Environment, Forest and Climate Change in consultation with the Forest Advisory Committee. The Central Empowered Committee (CEC), a judicially appointed committee that reviews forest clearance applications, was approached by two local activists who were part of the Niyamgiri Suraksha Samiti (Committee to Save Niyamgiri) to review the clearance being granted.

The CEC submitted a report to the Supreme Court in 2005, stating that the forest clearance should not be granted (CEC, 2005). The Forest Advisory Committee, which advises on the granting of such clearances, was asked to review the decision. This report by the CEC prompted the Norwegian Pension Fund to withdraw its funding on the grounds of violation of human rights. In 2006, the FRA was passed, and communities mobilized under this legal shield to claim their rights over forest land and sacred areas. While the claims for rights under the FRA were being initiated, a protracted social movement opposing the mine got underway. The Niyamgiri Suraksha

Samiti movement claimed that selling the land would be a violation of the right to FPIC of the forest-dwelling communities under international law namely, UNDRIP. The Dongria Kondh community was identified as a particularly vulnerable tribal group that had been living in and dependent on the Niyamgiri range of hills for generations. When the community learned about the government's plans to mine bauxite, the Niyamgiri Suraksha Samiti came into being as an initiative to protect their sacred hill though there was no recognition of the right to prior and informed consent, the forest-dwelling communities framed their protests on this basis.

In 2009, a circular by the Ministry of Environment, Forest and Climate Change recognized the right of consent of Gram Sabhas. This recognition formed the basis for the forest-dwelling community to approach the Supreme Court to claim retrospective violation. The Supreme Court admitted the case and, thus, began the process of judicial scrutiny. Domestic judicial scrutiny was accompanied by legal mobilization by the communities internationally.

A complaint was filed by the international non-governmental organization Survival International. The complaint was made through the National Contact Point (NCP) in the United Kingdom, which deals with violations of Organisation for Economic Co-operation and Development (OECD) guidelines for multinational enterprises.[12] In the event of a violation, the NCP can offer to bring together the company and complainant to engage in a process of mediation. Such an offer was not accepted by Vedanta. The NCP of the United Kingdom held that Vedanta Resources was in violation of the guidelines with respect to the lack of adequate and timely consultation (*Survival International v Vedanta Resources Pvt Ltd*, 2008). The filing of this complaint was also accompanied by informal modes of legal mobilization, in which the leader of Niyamgiri Suraksha Samiti protested outside of the annual general meeting of Vedanta in London.

As a domestic response to this international attention, a process was initiated to review the clearances granted to Vedanta. A committee headed by Dr Usha Ramanathan, an independent legal researcher and advocate, visited the site and reported that the clearance had been granted without implementing the FRA (Forest Advisory Committee, 2010). This was followed by a request to review the decision of the Forest Advisory Committee to grant a forest clearance (Forest Advisory Committee, 2010, 121). The committee produced a report under the chairmanship of Naresh Chandra Saxena, a retired Indian bureaucrat, which held that the FRA had not been adequately implemented and that if this project was granted clearance it would gravely harm the Indigenous Dongria Kondh people, in addition to the Dalit communities living in the area (MoEFCC, no date).

The Supreme Court case considered these multiple legal violations and recognized the authority of the Gram Sabha to decide on the issue through a vote, with emphasis on the right of communities to freedom of religion under Article 25 of the Constitution.[13] Twelve Gram Sabhas were recognized, though many more hamlets and villages were impacted by the mining activities. The Gram Sabhas voted against the mine and their decision was upheld by the government. In its judgement, the Supreme Court referred to the decision by the UK NCP as evidence of a violation of international legal standards, apart from the relevant sections of UNDRIP.

In response to the harsh criticism it received, Vedanta changed its internal stakeholder management policies to include the requirement of FPIC (Vedanta Resources Limited, 2013). Though the granting of the forest clearance initially was in violation of the sustainable development framework put forth by the Indian Bureau of Mines, this framework was stated to be a guiding document for decision-making and, therefore, not binding.

This narrative of legal mobilization reveals the engagement of social movements with legal pluralism that can be harnessed for the recognition of FPIC. As legal recourse was being pursued domestically, it was accompanied by legal action internationally. This kind of simultaneous challenge within the regulatory pluriverse sometimes enables the collusion between the state and business actors to be broken. Because legal mobilization within India is often undertaken in the tight frame of the pro-business development model and assertion of eminent domain, legal action outside its borders potentially enables a way to hold actors accountable. It is this leap outside the tight frame that is discussed next as the mode of jurisdictional leapfrogging.

Jurisdictional leapfrogging

We live in a time of porous legality or legal porosity, multiple networks of legal orders forcing us to constant transitions and trespassing. Our legal life is constituted by an intersection of different legal orders, that is, by inter-legality.

(Santos, 2002, 437)

Santos (2002) captures the postcolonial legal landscape as one of inter-legality, where multiple sources of law interact, coexist or conflict. Different legal spaces are superimposed on one another to create a diverse web of laws at different levels regulating diverse legal issues. The regulatory pluriverse of the Vedanta case illustrates this inter-legality. The superimposition of legal spaces offers social movements an opportunity to approach multiple legal fora to contest a legal violation or make a legal claim. It is this access to diverse fora and laws that I term jurisdictional leapfrogging.

The act of leapfrogging jurisdictions is referred to as externalization by Keck and Sikkink where external institutions are challenged to intervene in domestic affairs, as in the advocacy coalitions through which Brazilian rubber tappers and non-governmental organizations worked with northern environmental groups and put pressure on the World Bank to hold the Brazilian government to account for development projects in the Amazonian rainforests (Keck and Sikkink, 1998).

This strategy of legal mobilization is evident in instances where legal violations are committed by multinational corporations or where international investment is involved. The developmental state within which legal mobilization against mining takes place makes it difficult for the recognition of the right of consent. By placing the complaint before a legal forum that may not be embedded within the demands of the developmental state, jurisdictional leapfrogging offers forest-dwelling communities an avenue by which to challenge the legal violation in a potentially more impartial setting.

Legal mobilization, jurisdictional leapfrogging and
domestic-level decision-making in the Vedanta case

Jurisdictional leapfrogging shapes domestic-level decision-making when external institutions take on the role of intervening in domestic affairs. This was seen clearly in the Vedanta case. As Randeria (2003) states, international organizations like the World Bank and OECD introduce into the national legal arena concepts and principles which may considered proto-law: they do not have the formal status of law but in practice often obtain the same degree of obligation (Keck and Sikkink, 1998).

The proto-law, soft law and hard law concoction generates a scenario where the dispute is contended in various fora, using a varied legal vocabulary. For instance, the domestic legal dispute of Vedanta was framed as a violation of consent of the local village assembly. This takes on the form of lack of adequate and timely consultation when a complaint is filed before the NCP, in accordance with the multinational enterprise guidelines. The dispute is made to fit within these legal criteria in an effort to secure the rights of local communities to their lands and resources.

Randeria (2003) states that this form of legal and regulatory pluralism is linked on the one hand to the transnationalization of law and on the other hand to the simultaneous operation of multiple transnational norms, without their incorporation into domestic law. These international norms may not necessarily be diffused into domestic law but they add another layer of scrutiny in the implementation of domestic legal standards. The analysis of the Vedanta case is indicative of this phenomenon. The diffusion of norms, while coming from larger international legal instruments like the UNDRIP and UN International Labour Organization Convention 107 (Indigenous and Tribal Populations Convention), which are referred to in the Vedanta judgement of the Supreme Court, may not be occurring with respect to soft law instruments with an enforcement mechanism like the NCP.

Conclusion

The regulatory environment of FPIC in India is embedded in a political economy where the state prioritizes economic growth through mining. The desire for ease of doing business in the recent past has been accompanied by legislative attempts to dilute the right of consent. Land is being made available for mining by asserting the Doctrine of Eminent Domain. The enforcement of FPIC takes place in this formidable frame of a pro-business development model. As this chapter illustrates, this frame can sometimes be broken by the strategy of legal mobilization of jurisdictional leapfrogging.

The regulatory environment, as discussed in the chapter, has been influenced by the recent liberalization of the mining sector and the strategies of legal mobilization. This has brought about a form of regulatory bias, where there is better enforcement of the violation of FPIC in the case of multinational enterprises and the involvement of international investment than there is regarding state-owned companies. As the entrance of multinational enterprises in the mining sector increases with the expansion of coal mining, the contours of the regulatory environment will be reshaped. The enforcement of FPIC can be effectively implemented by leapfrogging jurisdictions, yet this may not occur in all cases. The structural win-lose will remain as long as the assertion of FPIC takes place within a pro-business development model.

As I was completing this chapter, a news article titled "Odisha's Niyamgiri Hills – and its people – are still under threat" captured the ongoing dilemma that configures this visible fault line of development (Singh, 2018). The people of Niyamgiri were safe for the time being, the article explained, but not for long, because the aluminium refinery had applied for permission to expand its operation. Members of the Niyamgiri Suraksha Samiti were arrested and accused of being part of the armed left-wing movement. The right of consent, once asserted and recognized, does not erase the fault line – it re-emerges with new threats and actors. Jurisdictional leapfrogging provides some respite for movements like Niyamgiri, but at times, the best it can do is delay extraction or check its excesses, as seen in the case against Coal India. It may not resolve this fault line entirely.

Notes

1 In this chapter, the terms free, prior and informed consent (FPIC) and right of consent are used interchangeably to refer to the right of the village assembly, or Gram Sabha.
2 Economic nationalism refers to a highly interventionist state with an import substitution policy.
3 The structural adjustment program was part of the credit conditionality imposed by the International Monetary Fund.
4 Ease of Doing Business rankings are part of the economic reforms put forth by the World Bank as part of its Doing Business indicator (https://www.doingbusiness.org/en/doingbusiness).
5 The Constitution of India has created a state list and a central list where division of legislative subjects demarcates the jurisdiction between the federal and subnational units.
6 The 73rd constitutional amendment (1992) established a three-tier local self-governance model to enable decentralization of decision making.
7 Scheduled areas are areas demarcated within Schedule V of the Indian Constitution and are protected from alienation to non-tribal communities.

8 "Public purpose," as defined in Section 2 of the Land Acquisition Act 1894, was all-encompassing and enabled acquisition of large tracts of land.
9 Environment Impact Assessment Notification 1994.
10 The new Land Acquisition Act requires the consent of 80 per cent of affected families. Affected families are described as landowners who have recognized rights to land and who are likely to lose their livelihood due to land acquisition. The Forest Rights Act 2006 requires that consent be obtained from the village assembly of the villages that have rights over forest land.
11 FPIC as interpreted, for instance, within the company in accordance to its own policies varies from the bureaucratic interpretation.
12 National Contact Points are set in OECD member countries to investigate and adjudicate on complaints brought under the OECD Guidelines for Multinational Enterprises.
13 Article 25 refers to the fundamental right to practice any religion under the Indian Constitution.

References

Bajpaee, C. (2016) "Modi, India and the emerging global economic order," *Journal of Asian Public Policy*, 9(2), 198–210.

Bera, S. (2013) "Life and times of Lado Sikaka" (blog), *Down to Earth*, 4 September. Available at: http://www.downtoearth.org.in/blog/life-and-times-of-lado-sikaka-42028 (Accessed: 4 May 2018).

Berman, P.S. (2012) *Global Legal Pluralism: A Jurisprudence of Law Beyond Borders*. Cambridge: Cambridge University Press.

Caldentey, E.P. (2008) "The concept and evolution of the developmental state," *International Journal of Political Economy*, 37(3), 27–53.

Cabinet Committee on Economic Affairs (2018) Cabinet Committee on Economic Affairs Resolution to Privatise Coal. Available at: http://www.pib.nic.in/PressReleaseIframePage.aspx?PRID=1521019 (Accessed: May 4, 2018).

CEC (Central Empowered Committee) (2005) *Report in IA no. 1324 regarding the Alumina Refinery Plant being set up by M/S Vedanta Alumna* [sic] *Limited*. Available at: http://assets.survivalinternational.org/static/files/behindthelies/CEC_report_smaller.pdf (Accessed: May 4, 2018).

Dash, M. (2016) "Rights-based legislation in practice: A view from Southern Odisha" in Nielsen, K.B. and Nielsen, A.G. (eds.), *Social Movements and the State in India*. London: Palgrave Macmillan, 161–183.

Forest Advisory Committee (2010) "FAC Draft." Available at: http://indiaenvironmentportal.org.in/files/rpt_usha.pdf (Accessed: 4 May 2018).

Harvey, D. (2013) *The New Imperialism*. Oxford: Oxford University Press.

Jebaraj, P. (2013) "Clearance for linear projects will not need Gram Sabha consent," *The Hindu (Chennai)*, 16 February.

Keck, M.S. and Sikkink, K. (1998) *Activists Beyond Borders: Advocacy Networks in International Politics*. Ithaca, NY: Cornell University Press

Kodiveri, A. (2017) "Changing terrain of environmental citizenship in India's forests," *Socio-Legal Review*, 12(2), 74–104.

Kohli, A. (2010) *Democracy and Development in India: From Socialism to Pro-Business*. Delhi: Oxford University Press.

Kohli, K. and Gupta, D. (2016) *Mapping Dilutions in a Central Law*. Working Paper Series, Centre for Policy Research, New Delhi.

Kohli, K. and Menon, M. (eds.) (2016). *Business Interests and the Environmental Crisis*. New Delhi: SAGE India.

Levien, M. (2015) "From primitive accumulation to regimes of dispossession: Six theses on India's land question," *Economic and Political Weekly*, 50(22), 146–157.

Make in India – Direct Foreign Investment (no date). Available at: http://www.makeinindia.com/policy/foreign-direct-investment (Accessed: 4 May 2018).

Misra, S.N. and Ghadai, S.K. (2016) "Make in India – The manufacturing conundrum," *PEOPLE: International Journal of Social Sciences*, 2(1), 1–8.

MoEFCC (Ministry of Environment, Forest and Climate Change) (no date) *Report of the Four Member Committee to Investigate the Proposal Submitted by Odisha Mining Corporation to Mine Niyamgiri*. Available at: http://envfor.nic.in/sites/default/files/Saxena_Vedanta-1.pdf (Accessed: 4 May 2018).

Orissa Mining Corporation v Ministry of Environment and Forest and Others (2013) 6 S.C.R. 881.

Panagariya, A. (2015) "Land acquisition: Rationale and the way forward." NITI Aayog, 3 August. Available at: http://niti.gov.in/writereaddata/files/document_publication/NITIBlog5_VC.pdf (Accessed: 4 May 2018).

Panda, J. (2016) "Narendra Modi and his mode of governance," *Journal of Asian Public Policy*, 9(2), 95–97.

Ramesh, J. and Khan, M.A. (2015) *Legislating for Justice*. Oxford: Oxford University Press.

Randeria, S. (2003) "Cunning states and unaccountable international institutions: Legal plurality, social movements and rights to local communities to common property resources," *European Journal of Sociology*, 44(1), 27–60.

Rodríguez-Garavito, C. (2010) "Ethnicity.gov: Global Governance, Indigenous peoples and the right to prior consultation in social minefields' (2010)," *Indiana Journal of Global Legal Studies*, 18(1), 263–305.

Samatha v State of Andhra Pradesh and Others Appeal (civil) (1997) 4601–02.

Santos, B.D.S. (2002) *Toward a New Legal Common Sense: Law, Globalization, and Emancipation*. London: Butterworths.

Santos, B.D.S. and Rodríguez-Garavito, C. (eds.) (2005) *Law and Globalisation from Below: Towards a Cosmopolitan Legality*. Cambridge: Cambridge University Press.

Singh, V. (2018) "Odisha's Niyamgiri Hills – and its people – are still under threat," *The Wire*, 23 April. Available at: https://thewire.in/rights/odishas-niyamgiri-hills-and-its-people-are-still-under-threat (Accessed: 4 May 2018).

Sinha, A. (2011) "An institutional perspective on the post-liberalization state in India" in Gupta, A. and Sivaramakrishnan, K. (eds.), *The State in India after Liberalization*. New York, NY: Routledge, 49–68.

Sundar, N. (ed.) (2009). *Legal Grounds: Natural Resources, Identity and the Law in Jharkhand*. New Delhi: Oxford University Press.

Survival International v Vedanta Resources Pvt Ltd (2008). Available at: https://www.oecdwatch.org/cases/Case_165 (Accessed: 4 May 2018).

Tamanaha, B.Z. (2007) "Understanding legal pluralism: Past to present, local to global," *Sydney Law Review*, 30(3), 375–411.

Vedanta Resources Limited (2013) *Technical Guidance Document*. Available at: http://sustainabledevelopment.vedantaresources.com/content/dam/vedanta/corporate/documents/Scott%20Wilson%20Reports/ScottWilson_Jan2013.pdf.pdf   (Accessed: 4 May 2018).

4 Realizing sustainable development and water justice through procedural justice

Gayathri D. Naik

Introduction

Water is crucial for life and is a core component of environmental systems and sustainable development (Chaturvedi, 2012). Because human life depends on water, the right to water is indispensable to realizing all other basic rights. Every human being is entitled to adequate supplies of safe, clean and affordable water. Access to water has inspired cooperation as well as conflict among nations, and conflict can arise from both natural and anthropogenic factors. Natural factors such as drought, flooding and soil erosion can result in a lack of adequate water resources in terms of both quality and quantity. Inequity in distribution, anthropogenic pollution and competition over access are factors resulting from unsustainable water consumption and management, which in turn are due to increasing demands of a burgeoning population (ACF-International Framework, 2008). The influence of these anthropogenic factors is complex in groundwater access in India, which is governed by the land–water nexus wherein groundwater is considered as a chattel attached to land, with access rights confined to the landowner (Moench, 1992).

These factors often result in distributive injustices (Miller, 1999), which can be magnified by procedural injustices where the stakeholders are denied access to decision-making. This chapter investigates such procedural inequities in groundwater access in India. Procedural justice includes participatory decision-making, which is preceded by the right to adequate information and succeeded by access to justice. Groundwater laws in India have failed to ensure participatory justice for users of the resource, although a recent bill attempts to rectify this apparent injustice. In this chapter, I examine existing groundwater laws and highlight the need for procedural justice. Beginning with a brief introduction on international and Indian perspectives on the human right to water, I then discuss procedural justice and its three basic elements. The human right to water is reviewed in light of the groundwater situation in India; groundwater laws are analysed and suggestions for reform are highlighted.

DOI: 10.4324/9781003141228-4

Human right to water: international and national perspectives

Recognition of water as a human right has been hotly debated, with several scholars arguing the full extent and content of the right (Anon, 2007; Cahill, 2005; Gleick, 1999; McCaffrey, 1992, 2005). The United Nations Committee on Economic, Social and Cultural Rights (2002) General Comment 15 clarified matters to some degree, yet this comment, which is not binding on states, left the debate open (Clark, 2017; Cullet, 2013; Tully, 2005). Globally, 748 million people lack access to improved water sources, while billions lack safe drinking water [WWAP (United Nations World Water Assessment Programme), 2015]. Lack of access to sufficient quality and quantity of water for millions around the world deepens the global discussion on the right to water, the normative content of this right and its implementation at domestic levels. Despite its significance, neither the Universal Declaration of Human Rights (UNGA Res 217A, 10 December 1948) nor the International Covenants on Civil and Political Rights (adopted 16 December 1966, entry into force 23 March 1976) and Economic, Social and Cultural Rights (adopted 16 December 1966, entry into force 3 January 1976) (hereinafter International Bill of Rights) mention the right to water as a human right. Thielbörger (2015, 227) observes that omission of the right to water in the covenants could be "[e]ither considered as the expression of a deliberate silence expressing the states' (unspoken) consensus that there should not be such a thing as a human right to water; or, alternatively, as a negligent silence meaning that the human right to water was simply forgotten at the time of drafting the two Covenants."

Since the development of international human rights law, three instruments have explicitly mentioned the right to water as a human right. These are the Convention on the Elimination of All Forms of Discrimination against Women (18 December 1979, entered into force 3 September 1981) 1249 UNTS 13, art 14; the Convention on the Rights of the Child (adopted 20 November 1989, entered into force 2 September 1990) 1577 UNTS 3 art 24; and the Convention on the Rights of Persons with Disabilities (adopted 13 December 2006, entered into force 3 May 2008) 2515 UNTS 3, art 28. However, since these instruments are subject-specific and targeted only to the protection of women and children or persons with disabilities, the right to water as an individual right for all persons and binding on states is still absent in international human rights law (Olleta, 2010). Despite the absence of explicit recognition of the right to water in the International Bill of Rights or any other human rights treaties, the United Nations Committee on Economic Social and Cultural Rights has inferred the right to water from the International Covenant on Economic, Social and Cultural Rights.

In the United Nations Committee on Economic, Social and Cultural Rights General Comment 15 (2002), the right to water is defined as entitling everyone to sufficient, safe, acceptable, physically accessible and affordable water. The right to water contains both freedoms and entitlements. Freedoms include the right to maintain undisturbed access to water supply sources, including the right to be free from arbitrary disconnections, while entitlements point to the right to a system of supply and management that ensures equality of opportunity to enjoy this human right to water. General Comment 15 is a significant development in international law recognizing the human right to water. Yet this development has not achieved implementation at the domestic level on account of its non-binding nature.

India has been a forerunner in recognizing the fundamental nature of the right to water. Jurisprudential development of the right to water as a fundamental right is independent of international recognitions and developments. In India, the Constitution does not recognize the right to water as a fundamental right, but it is the judiciary that has carved out this right from the right to life. See, for example, *Subhash Kumar v State of Bihar* (1991) AIR SC 420; *Hamid Khan v State of Madhya Pradesh* (1997) AIR MP 191; *Vishala Kochikudivella Samarkhana Samiti v State of Kerala* (2006)(1) KLT 919; and *Narmada Bachavo Andolan v Union of India* (2000) AIR SC 3751. In the *Subhash Kumar* case, it was held that the "right to live is a fundamental right under Article 21 of the Constitution and it includes the right to enjoyment of pollution free water and air for full enjoyment of life." This case law points to the fact that in India, the right to water, though a fundamental right recognized by the courts, is recognized as a prerequisite for enjoying the right to life rather than acknowledging it as a stand-alone right. The courts have not elaborated on the content of this right, leaving it to the legislature, which has not yet enacted a law recognizing the right to water as a fundamental right.

According to Cullet (2013), the absence of constitutional recognition has not stopped the development of the right to water in various directions and contexts. He points out that courts have been at the forefront of explicit discussion of this right, much state legislation has aimed at the realization of the right and policy instruments by the union government regarding drinking water have contributed to the realization of the right in rural areas. Unfortunately, water justice issues persist with millions of people unable to access safe, adequate and affordable water. Court judgements and various water supply policies are often framed in broad terms, referring to water resources, and they tend to implicitly focus on the regulation and management of surface water without emphasizing the common-pool nature of these resources (Ohdedhar, 2019). Groundwater is left inadequately regulated and under-managed, amplifying the water crisis discussed below.

Realizing the right to water through procedural justice

As discussed in the previous section, the right to water is derived from the right to life. It is always discussed as a right required to exercise other human rights like the right to life, the right to food and the right to health. This description of the interconnected nature of the right to water reflects the anthropocentric orientation of the right as it currently stands. The rights of nature and the rights of rivers and other water bodies to have a natural flow of water are not adequately addressed in the water rights discourse. The case is the same in Indian jurisprudence on water rights. A demand-driven approach to water distribution often neglects the supply side and the basic needs of communities and people. In addition, denial of opportunities for stakeholders to be involved in decision-making also results in injustice.

Environmental justice aims to address issues of inequitable benefit and burden sharing around environmental goods (Bryant, 1995; Bullard, 1993; Eckersley, 1992; Mutz, Bryner and Kenney, 2001). It emphasizes an unfair distribution of environmental burdens on poorer communities that remain major victims of environmental harms without having contributed to these harms (Adealo, 2000; Nollkaemper, 2009). Environmental justice has distributive, recognitional and procedural elements embedded in it (Schlosberg, 2007). While distributive justice focuses on substantive questions, recognitional and procedural justice pertain to process questions and matters of implementation (Sambo, 2012).

Procedural justice, an important element of water justice, is concerned with the participation in decision-making processes for projects that may create environmental harm or judicial proceedings to challenge administrative decisions and is, thus, linked to fairness, transparency and accountability (Razzaque, 2009). Public participation denotes informing, consulting and participating in environmental issues like planning, management and impact assessment of significant and large sector projects (Clark, 1994). The significance of public participation has been upheld in the international environmental law through the Rio Declaration on Environment and Development (UN Doc. A/CONF.151/26, Vol. I, 1992), Principle 10, which calls upon states to provide their citizens appropriate opportunities to participate in environmental decision-making. States shall provide effective access to judicial and administrative proceedings, including redress and remedy. Though the Rio Declaration is a non-binding soft instrument, it has influenced tremendously the development of public participation in decision-making in several other international treaty mechanisms, such as the Aarhus Convention of 1998 (United Nations Economic Commission for Europe Convention on Access to Information, Public Participation in Decision-Making and Access to Justice in Environmental Matters, 1999, 38 International Legal Materials 515). Despite being regional in application, the Aarhus Convention is significant in ensuring procedural justice through public participation in environmental matters. This convention is the first treaty that prioritizes rights of the public with respect to the environment. It provides for the right to receive environmental information from public authorities (access to information); the right to participate in environmental decision-making (public participation in environmental decision-making); and the right to access judicial mechanisms to challenge the public decision on environment (access to justice). The principles and rights enshrined in this convention are relevant to the development of environmental and water justice in India. They could be applicable in the case of transboundary water sharing and water disputes when people have not received information, been granted the opportunity to participate in decision-making or given access to courts for redressal of any harm to their human rights or environment.

Participatory rights are incomplete without adequate recognition of the stakeholders involved in any process (Mbano-Mweso, 2015). Many environmental justice scholars argue that recognition of the communities affected by environmental harm is a prerequisite for ensuring fair participation and procedural and distributive justice (e.g., Schlosberg, 2004) – and this is clearly applicable to the water sector. Applying recognitional justice to the water sector would strengthen procedural justice through maximizing participation of all stakeholders. Doing this would help bring to light stakeholder concerns along with ways to address the concerns. This is particularly significant in decision-making about water, such as conducting environmental impact assessments of large-scale development projects on rivers. In cases like that, displaced people like Adivasis (often viewed as the original inhabitants of India), Dalits, women and the vulnerable should be recognized with due respect of their rights to ensure their effective participation.

In the water sector in India, major legislation related to irrigation comes from the pre-independence era, and this law concentrated power with the state while not providing for the participation of farmers. After independence, the irrigation sector witnessed changes with the strengthening of participatory irrigation management through state legislation (Cullet, 2015; Madhav, 2010). The legal framework for groundwater has remained the same since the colonial period, with the only stakeholder being the landowner, as per the existing interpretation of the application of the Easement Act to groundwater rights (Koonan, 2013). In water management generally, the literature suggests that meaningful opportunities for public participation would provide an "informed public, legitimate remediation plans, more efficient implementation of measures, and a reduction of conflict among stakeholders" (Jonsson, 2005, 495).

Water policies in India have public participation embedded in them. In the absence of a codified law on groundwater, other than the Water Act 1974 (which bases its priorities on prevention and abatement of water pollution rather than water rights), these policies are of fundamental importance. They are critical for understanding developments in the water sector and they form the foundations for all central government water supply programmes. Since the first water policy of 1987, subsequent policies have tried to bring stakeholders' voices into the water sector. The latest policy, from 2012, highlights community participation as a part of good governance in the management of water projects and services (National Water Policy 2012, sections 1 and 12).

To ensure participation, proper and transparent information is essential, and for community members, the Right to Information Act 2005 is an important tool for gaining access to information. Redressal of grievances is ensured through a responsible judiciary and constitutional protections (see particularly Articles 14, 15, 17, 21, 32 and 226, which are used for availing remedies in the water sector). Part III of the Constitution is the basis for all environmental activism in India. Public interest litigation has helped millions of people in cases dealing with water quality and supplies. It has also helped to make the state responsible and accountable to create statutory bodies. The case of *M.C. Mehta v Union of India* (1987) SCR (1) 819 is a good example. The Central Groundwater Authority was constituted under section 3(3) of the Environment (Protection) Act because of specific directions given by the Supreme Court in that case to create a body to deal with groundwater extraction problems (Aguilar, 2011). Hence, procedural justice has become essential for complementing distributive justice in the context of the constitutionally enshrined right to water for every individual without any discrimination.

Groundwater rights in India: an intersection of land–water rights

Groundwater plays a major role in sustaining livelihood security, especially in agricultural economies (Shankar, Kulkarni and Krishnan, 2011). It constitutes the main source of drinking water in many parts of India (Moench, 1992). The dependence on groundwater has been increasing at an alarming rate, owing to its role in ensuring food security and clean drinking water (Cullet, 2014). The extraction rate of this "invisible, non-stationary, 'fugitive' resource" (Kulkarni and Shankar, 2009) in many Indian states has exceeded annual recharge, resulting in the lowering of the water table (Planning Commission of India, 2007, 14).

Water law in India has not been codified (Cullet, 2015). The existing legal framework regulating groundwater in India is complicated, encompassing constitutional provisions, various states' irrigation and land laws, state and central legislation, judicial decisions, common law principles and customary practices (Aguilar, 2011). The current legal framework, which is based on "outdated, inequitable and unsustainable principles" that link groundwater to landownership (Cullet, 2009), has not addressed the growing water requirements of the country (Koonan, 2019) and failed to advance water justice in India.

Relevant common law principles were developed at a time when groundwater use was marginal and limited to minor use of wells and springs (Cullet, 2011). The principles for the regulation of groundwater in India were developed by the courts in England in the nineteenth century. These include *Acton v Blundell* (1843) [1843] 152 ER 1223, *Chesmore v Richards* (1860) 7 H.C.L Cas 349 [1860] and *English v Metropolitan Water Board* (1907) (1907) 1 K.B 588324 (Cullet, 2011; Koonan, 2010). Those principles, suited for a different climatic condition and developed at a time when groundwater–surface water hydrology was not well developed, are still in use in many parts of India without any formal legislative changes (Cullet, 2011).

The pre-constitutional era legislation also followed the same approach in regard to groundwater as a part of landownership. The Indian Easement Act 1882 provides evidence of the influence of the common law in determining groundwater rights. Groundwater is considered as a chattel attached to land property, with rights to groundwater being vested in the landowner (Singh, 1991b, 1992) on the basis of the ad coelum principle (meaning that the landowner owns everything beneath and above his or her land) (Jain, 1981; Puthucherril, 2009). By virtue of section 7(b) of the Easement Act, every owner of land has an easement right to collect and dispose of, within his or her own limits, all waters under the land which do not pass in a defined channel. As per the provisions of the Transfer of Property Act 1882 and the Land Acquisition Act 1894, groundwater extraction rights cannot be separated from landownership, and transfer of such rights can only be done when there is ownership in such land (Moench, 1998; Singh, 1992). Groundwater rights are, therefore, considered by some as being linked to landownership as an easement right (Garduño et al., 2011; Saleth, 1994).

However, Vani (2009) observes that the right to groundwater is not an easement right, but a right attached to land that may be restricted by easements. An easement, which requires a dominant and a servient heritage, cannot coexist with the ownership of land in a single person. In other words, groundwater rights cannot be governed by easement right, since for an easement right to be implemented, two different lands are required. In the case of groundwater, we speak about the water only beneath one's land, which makes it difficult to apply a principle like an easement right to groundwater. Similarly, an easement right would enable the right holder innumerable legal actions for remedy whenever the act of any person on his or her land resulted in the diminution of water in the adjacent land, which is not available in the case of groundwater issues (Vani, 2009).

The principles of landownership and easement rights, which were developed when knowledge of basic hydrology was not well developed, have proven inapplicable in an era when advances in science and technology have made it possible to determine various characteristics of groundwater flows and their interrelation with surface water, thus calling for a holistic treatment of whole water systems. Moreover, the human rights jurisprudence that declared the right to water as a fundamental human right makes the private ownership of groundwater untenable.

Depletion of groundwater, the primary source of water for drinking, irrigation and domestic uses in India, is considered as the first signal or indication of an impending water scarcity (Shankar, Kulkarni and Krishnan, 2011). In a country where the main source of irrigation is groundwater (Mukherji, Rawat and Shah, 2013), linking landownership and access to groundwater rights leads to landless people and tribes with collective rights over land being deprived of a fundamental human right as well as opportunities to pursue their livelihoods (Singh, 1991b).

The introduction of hybrid rice varieties and extension of paddy culti-vation to non-paddy areas as part of the green revolution augmented the exploitation of groundwater resources with tube well technology in areas like Punjab, where there was insufficient supply and inefficient management of canal irrigation (Singh, 1991a). The cropping pattern in many groundwa-ter-irrigated areas depends on the easy availability of groundwater, leading to water-intensive agriculture (Planning Commission of India, 2007). This is further compounded by the power subsidies and horsepower tariffs for agri-culture provided by the states, which encourage overexploitation of ground-water (Koonan, 2010). This is especially true for many parts of peninsular India, where the proliferation and energization of wells have made possible excessive withdrawal of groundwater, leading to depletion of the resource.

Procedural justice in access to groundwater in India: an equity dimension

As a federal nation, India has a constitutional division of powers between the central and state governments. Under the constitutional scheme, water is included as Entry 17 in the state list, empowering the states to enact laws on the same (see Article 246, which regulates the lists under Schedule VII). However, this entry is subject to Entry 56 of the union list, which grants jurisdiction to the central government with respect to interstate rivers and river valleys to the extent to which such regulation and development is declared by Parliament by law to be expedient in the public interest. Article 248 empowers Parliament to enact laws on those matters not granted explic-itly to the states. Realizing the growing significance of groundwater in the country and the need to reform its existing legal principles, the central gov-ernment introduced a model bill to Regulate and Control the Development and Management of Groundwater (Model Bill 1970/2005) (Cullet, 2012). The bill retains the basic scheme adopted in 1970, though it underwent sub-sequent amendments in 1992, 1996 and 2005. Although water falls primar-ily under the jurisdiction of the states, this model bill is important because it reveals the position of the central government on the best approaches to address the emerging groundwater concerns in the country (Moench, 1994). Several states have followed the 2005 bill when enacting state legislation. The bill and subsequent state legislation are state-centred and grant powers and responsibilities to groundwater authorities without any provision for stake-holder participation. Furthermore, the bill and state laws fail to address the injustices faced by those without land, thus further violating their fundamen-tal right to water. Landless people, Dalits and women are the most affected by this and are without redress for over extraction of water from land adjacent to where they live and work. This threatens the justice and equality envisaged by the Constitution.

In the Model Groundwater (Sustainable Management) Bill from 2016, the central government tried to incorporate a decentralized approach to managing groundwater resources. The bill envisages a multi-tiered institutional framework involving subcommittees at the Gram Panchayat (village council) level, management committees at the block panchayat and municipal levels and district councils and advisory councils at the state level. It also envisages the application of social and environmental impact assessment and public consultation provisions to groundwater projects. Duties are corollary to rights. While the bill guarantees citizens the fundamental right to water, it also makes them duty-bound to protect the resource by including provisions regarding "duties of groundwater users" (Naik, 2018). Since this bill has not been accepted by the states in enacting or amending existing laws, the status quo has not yet changed.

Conclusion

In India, the right to water is fundamental but the content of the right and its mode of implementation have not been fully developed by the courts or legislatures. This gap is exacerbated in the case of groundwater because of the ongoing applicability of common law principles that tie groundwater rights to landownership (the land–water nexus). The Constitution of India envisages distributive justice, but this goal has been blocked in the case of groundwater. Only with a change to the existing legal framework can this be rectified to ensure equitable groundwater access for everyone. In addition, procedural justice should supplement this initiative by guaranteeing meaningful opportunities for public participation, supported by the right of access to information and succeeded by access to justice. Presently, no access to justice is available to someone whose neighbour drills the land and extracts water. Hence, public participation should be mandatory in the case of decisions regarding groundwater, as this is the main source of water for drinking and irrigation in the country. Procedural justice will not be complete unless it is linked to recognitional justice, which is required to ensure recognition of affected persons in any water-related projects or instances of water supply. In many cases, the voices of victims are not heard because there is no acknowledgement that they are affected. Their rights and culture (in cases like Adivasis) and their concerns should be duly recognized in order to ensure their participation in decision-making. This again mandates the need for information dissemination and access to information as the first step. Public participation will be complete once there is provision for access to justice for grievance redressal. The Right to Information Act 2005 and public interest litigation help support access to information and access to justice. However, public participation is still not successful in the country's water sector. Justice is the goal of the Constitution – to ensure this and establish an equitable welfare state, distributive, recognitional and procedural justice are essential. In the case of groundwater, these elements of environmental justice are urgently needed to protect, preserve and uphold intergenerational equity.

62   *Gayathri D. Naik*

References

ACF-International Framework (2008). *The Right to Water: Emergence, Definition, Current Situation and Stakeholder's Positions.* Available at: https://www.actioncontrelafaim.org/wp-content/uploads/2018/01/acf-right_to_water-janv2008.pdf (Accessed 2020).

Acton v Blundell (1843) [1843] 152 ER 1223.

Adealo, F.O. (2000) "Cross-national environmental injustice and human rights issues," *American Behavioural Scientist,* 43(4), 686–706.

Aguilar, D. (2011) "Groundwater reform in India: An equity and sustainability dilemma," *Texas International Law Journal,* 46(3), 623–653.

Anon. (2007) "What price for the priceless? Implementing the justiciability of the right to water," *Harvard Law Review,* 120(4), 1067–1088.

Bryant, B. (ed.) (1995) *Environmental Justice: Issues, Policies, and Solutions.* Washington, DC: Island Press.

Bullard, R. (ed.) (1993) *Confronting Racism: Voices from the Grassroots.* Boston, MA: South End Press.

Cahill, A. (2005) "'The human right to water – A right of unique status': The legal status and normative content of the right to water," *International Journal of Human Rights,* 9(3), 389–410.

Chaturvedi, M.C. (2012) *India's Waters: Advances in Development and Management.* Boca Raton, FL: CRC Press.

Chesmore v Richards. (1860) 7 H.C.L Cas 349 [1860].

Clark, B. (1994) "Improving public participation in environmental impact assessment," *Built Environment (1978–),* 20(4), 294–308.

Clark, C. (2017) "Of what use is a deradicalized human right to water?" *Human Rights Law Review,* 17(2), 231–260.

Cullet, P. (2009) *Water Law, Poverty, and Development: Water Sector Reforms in India.* Oxford: Oxford University Press.

Cullet, P. (2011) *Groundwater Regulation – Need for Further Reforms.* IELRC Working Paper.

Cullet, P. (2012) "The groundwater model bill: Rethinking regulation for the primary source of water," *Economic and Political Weekly,* 47(45), 40–47.

Cullet, P. (2013) "Right to water in India – Plugging conceptual and practical gaps," *The International Journal of Human Rights,* 17(1), 56–78.

Cullet, P. (2014) "Groundwater law in India: Towards a framework ensuring equitable access and aquifer protection," *Journal of Environmental Law,* 26(1), 55–81.

Cullet, P. (2015) "Water regulation and public participation in the Indian context" in Tignino, M. and Sangbana, K. (eds.), *Public Participation and Water Resources Management: Where Do We Stand in International Law?* Paris: UNESCO.

Eckersley, R. (1992) *Environmentalism and Political Theory: Towards an Ecocentric Approach.* London: UCL Press.

English v Metropolitan Water Board. (1907) (1907) 1 K.B 588324.

Garduño, H., Romani, S., Sengupta, B., Tuinhof, A. and Davis, R. (2011) *India Groundwater Governance Case Study.* World Bank Water Papers, Washington, DC: World Bank.

Gleick, P. (1999) "The human right to water," *Water Policy,* 1(5), 487–503.

Hamid Khan v State of Madhya Pradesh (1997) AIR 1997 MP 191.

Jain, S.N. (1981) "Legal aspects of groundwater management," *Journal of Indian Law Institute,* 23(2), 181–189.

Jonsson, A. (2005) "Public participation in water resources management: Stakeholder voices on degree, scale, potential, and methods in future water management," *AMBIO*, 34(7), 495–500.

Koonan, S. (2010) "Legal regime governing groundwater" in Cullet, P., Gowlland-Gualtieri, A., Madhav, R. and Ramanathan, U. (eds.), *Water Law for the Twenty-First Century: National and International Aspects of Water Law Reforms in India*. Abingdon: Routledge, 182–204.

Koonan, S. (2013) *Groundwater Legal Regime in India: Towards Ensuring Equity and Human Rights*. SOAS Law Working Papers, 1. Available at: https://eprints.soas.ac.uk/23103/1/13-1%20Koonan-Ground%20water%20legal%20regime%20in%20India.pdf.

Koonan, S. (2019) "Law regarding groundwater" (blog), *Law Umbrella*, 5 January. Available at: https://lawumbrella.wordpress.com/2019/01/05/law-regarding-groundwater/.

Kulkarni, H. and Shankar, P.S.V. (2009) "Groundwater: Towards an aquifer management framework," *Economic and Political Weekly*, 44(6), 13–15, 17.

M.C. Mehta v Union of India (1987) (1987) 11 SCC 312.

Madhav, R. (2010) "Law and policy reforms for irrigation" in Cullet, P., Gowlland-Gualtieri, A., Madhav, R. and Ramanathan, U. (eds.), *Water Law for the Twenty-First Century: National and International Aspects of Water Law Reforms in India*. Abingdon: Routledge, 205–234.

Mbano-Mweso, N.N. (2015) *Realising the Human Right to Water in Malawi through Community Participation*. LLD thesis, University of the Western Cape. Available at: https://core.ac.uk/download/pdf/58916199.pdf.

McCaffrey, S. (1992) "A human right to water: Domestic and international implications," *Georgetown International Environmental Law Review*, 5(1), 1–24.

McCaffrey, S. (2005) "The basic right to water" in Weiss, E.B., Chazournes, L.B. de and Bernasconi-Osterwalder, N. (eds.), *Fresh Water and International Economic Law*. Oxford: Oxford University Press, 93–115.

Miller, D. (1999) *Principles of Social Justice*. Cambridge, MA: Harvard University Press.

Moench, M. (1992) "Drawing down the buffer: Science and politics of ground water management in India," *Economic and Political Weekly*, 27(13), A7–A14.

Moench, M. (1994) "Approaches to ground water management: To control or enable?" *Economic and Political Weekly*, 29(39), A135–A146.

Moench, M. (1998) "Allocating the common heritage: Debates over water rights and governance structures in India," *Economic and Political Weekly*, 33(26), A46–A53.

Mukherji, A., Rawat, S. and Shah, T. (2013) "Major insights from India's minor irrigation censuses: 1986-87 to 2006-07," *Economic and Political Weekly*, 48(26/27), 115–124.

Mutz, K.M., Bryner, G.C. and Kenney, D.S. (eds.) (2001) *Justice and Natural Resources: Concepts, Strategies and Applications*. Washington, DC: Island Press.

Naik, G.D. (2018) "Groundwater regulation in India: Applicability of public trust doctrine and right to participation in decision making to achieve right to water" in Rehman, J. and Shahid, A. (eds.), *Asian Yearbook of Human Rights and Humanitarian Law*, vol. 2. Leiden: Brill Nijhoff, 327–349.

Narmada Bachavo Andolan v Union of India (2000) AIR 2000 SC 3751.

Nollkaemper, A. (2009) "Sovereignty and environmental justice in international law" in Ebbesson, J. and Okowa, P. (eds.), *Environmental Law and Justice in Context*. Cambridge: Cambridge University Press, 253–269.

Ohdedhar, B. (2019) "The human right to water in India: In search of an alternative commons-based approach in the context of climate change" in Haller, T., Breu, De Moor, T., Rohr, C. and Znoj, H (eds.), *The Commons in a Global World: Global Connections and Local Responses*. Abingdon: Routledge, 475–492.

Olleta, A. (2010) "An overview of common trends in the water legislation of selected jurisdictions" in Cullet, P., Gowlland-Gualtieri, A., Madhav, R. and Ramanathan, U. (eds.), *Water Law for the Twenty-First Century: National and International Aspects of Water Law Reform in India*. Abingdon: Routledge, 11–35.

Planning Commission of India. (2007) *Report of the Expert Group on Groundwater Management and Ownership*. Government of India.

Puthucherril, T.G. (2009) "Riparianism in Indian water jurisprudence" in Iyer, R. (ed.), *Water and the Laws in India*. New Delhi: Sage Publications India, 97–133.

Razzaque, J. (2009) "Participatory rights in natural resource management: The role of communities in South Asia" in Ebbesson, J. and Okowa, P. (eds.), *Environmental Law and Justice in Context*. Cambridge: Cambridge University Press, 117–138.

Saleth, M. (1994) "Groundwater markets in India: A legal and institutional perspective," *Indian Economic Review*, 39(2), 157–176.

Sambo, P.T. (2012) A Conceptual Analysis of Environmental Justice Approaches: Procedural Environmental Justice in the EIA Process in South Africa and Zambia. PhD thesis, University of Manchester. Available at: https://www.research.manchester. ac.uk/portal/files/54523801/FULL_TEXT.PDF.

Schlosberg, D. (2004) "Reconceiving environmental justice: Global movements and political theories," *Environmental Politics*, 13(3), 517–540.

Schlosberg, D. (2007) *Defining Environmental Justice: Theories, Movements, and Nature*. Oxford: Oxford University Press.

Singh, S. (1991a) "Some aspects of groundwater balance in Punjab," *Economic and Political Weekly*, 26(52), A146–A155.

Singh, C. (1991b) *Water Rights and Principles of Water Resources Management*. Bombay: N.M. Tripathi.

Singh, C. (1992) "Water rights in India" in C. Singh (ed.), *Water Law in India*. Bombay: N.M. Tripathi, 8–31.

Shankar, P.S.V., Kulkarni, H. and Krishnan, S. (2011) "India's groundwater challenge and the way forward," *Economic and Political Weekly*, 46(2), 37–45.

Subhash Kumar v State of Bihar (1991) AIR 1991 SC 420.

Thielbörger, P. (2015) "Re-conceptualizing the human right to water: A pledge for a hybrid approach," *Human Rights Law Review*, 15(2), 225–249.

Tully, S. (2005) "A human right to access water? A critique of General Comment No. 15," *Netherlands Quarterly of Human Rights*, 23(1), 35–63.

United Nations Committee on Economic, Social and Cultural Rights (2002) *General Comment No. 15 on the Right to Water*, UN Doc. E/C.12/2002/11.

Vani, M.S. (2009) "Groundwater law in India: A new approach" in Iyer, R.R. (ed.), *Water and the Laws in India*. New Delhi: Sage Publications India, 435–474.

Vishala Kochikudivella Samarkhana Samiti v State of Kerala (2006) 2006(1) KLT 919.

WWAP (United Nations World Water Assessment Programme). (2015) *The United Nations World Water Development Report 2015: Water for a Sustainable World*. Paris: UNESCO.

5 The National Green Tribunal's response to the cause of tribals and fisherfolk

Armin Rosencranz, Avinash Mathews, Sangram Sinh Parab, Shubham Janghu and Suhaas Putta

Tribals

Amnesty International, in its report titled *When Land Is Lost, Do We Eat Coal?* (Amnesty International, 2016), observed that Indian authorities have continued to breach their obligations to protect the tribal communities affected by mining projects under way, *inter alia,* in Chhattisgarh, Jharkhand and Odisha. These state governments have failed to evaluate the impact of mining activity on the human rights of tribal peoples. Given such grave violations, it becomes the duty of the National Green Tribunal (NGT) to offer effective remedies so as to ensure distributive and restorative justice.

Background to tribals' forest rights and legislation in India

In 2006, Parliament enacted the Scheduled Tribes and Other Traditional Forest Dwellers (Recognition of Forest Rights) Act 2006 (also referred to as the Forest Rights Act or FRA). The essence of this act is to recognize various rights that are pivotal to both the lives and livelihoods of India's tribal population. These include rights to ownership of minor forest produce, water bodies, habitat of primitive tribal groups and the power to protect, conserve and manage community forest resources (Council for Social Development, 2010). Prior to the enactment of this statute, the above-mentioned rights were denied to these remote tribal communities. In essence, the FRA strives to correct such shortcomings and address grievances that were not addressed by 1990 guidelines issued by the Ministry of Environment and Forests (now the Ministry of Environment, Forest and Climate Change).

The inception of the FRA was a consequence of the historical deprivation of rights to the tribal and other forest communities (Springate-Baginski et al., 2009). During colonial rule in India, the British implemented legislative measures such as the Indian Forest Act 1864 with the sole intent of centralizing and restricting forest use for the purposes of commercial exploitation by clearing large tracts of flora and instead growing opium and indigo. This policy remained unchanged even after India became independent (Sarin, Singh, Sundar and Bhogal, 2003). Large areas of land were transferred to the forest department without paying any

DOI: 10.4324/9781003141228-5

heed to the rights of the tribal population. In cases where forests were notified as reserved or private, there was a significant watering down of rights available to forest-dependent people. For example, their public objection had no impact on the granting of an environmental clearance.

The enactment of the Wildlife (Protection) Act of 1972 and the Forest (Conservation) Act of 1980, in response to the increasing attention to the conservation of flora and fauna, has resulted in forest-dependent and tribal communities being labelled as encroachers on their own lands (Bose, 2010). Apart from conservation concerns, there is continuing pressure for development in India. Construction of hydroelectric power plants, mines and numerous other kinds of industrial projects has resulted in a further displacement and marginalization of tribal communities.

Between 1950 and 1990, an estimated 21.3 million forest dwellers were displaced. Forty per cent of these displacees belonged to the Scheduled Tribe category, which forms 8 per cent of the total population of India (Council for Social Development, 2010). Conservation and development in India have come at the expense of permanently displacing such communities. As a consequence of the protests by civil society, the Ministry of Tribal Affairs drafted a bill recognizing the rights of tribal and forest-dependent people, in contrast to the retrogressive approach of the Ministry of Environment, which had previously handled tribal affairs (Amnesty International, 2016).

A majority of stakeholders aligned themselves with one of two major groups – one favouring the idea of conservation by maintaining the status quo and the other wanting an overhaul of the rights of tribal people. The formation of two distinct groups created unease between the Ministry of Tribal Affairs and the Ministry of Environment, with powerful lobbies working on either side. It has been claimed that the conservation lobby was behind the launch of the "misinformation campaign" that garnered immense support from the general public (Springate-Baginski et al., 2009). For example, the conservation lobby argued that the proposed bill of rights, which vested certain rights in tribal communities, would result in a loss of 15 per cent of India's forest cover, while the tribal lobby strongly asserted that according to the bill, only 2 per cent of the lands would be under its purview. The mounting pressure from civil society resulted in the enactment of the FRA 2006 (Sarin and Springate-Baginski, 2010).

NGT response to the tribals' cause

The NGT is counted among one of the few institutions in India that holds public trust (Ramachandran, 2016). Since its inception in 2010, the tribunal judgements have set forth an exemplary model emphasizing the need to advance environmental jurisprudence. The approach of the NGT has been strict, leaving no stone unturned in holding large corporations, the Ministry of Environment and governments accountable for not paying heed to forest dwellers and tribal rights advocates.

Balance between development and environment

The recent trend and attitude of the NGT have reflected a greater emphasis on the protection of the environment along with an attentiveness to the needs and views of those affected by potential development projects. A recent decision of the NGT (*Paryawaran Sanrakshan Sangarsh Samiti Lippa v Union of India*, 2011), which demonstrates this outlook, created unease in the corporate sector. The decision attaches greater significance to local populations and Panchayati Raj (local government) institutions, including the Gram Sabhas (village assemblies) of tribal districts. After a seven-year struggle, the NGT granted a village settlement in Himachal Pradesh the right to decide upon its preferred route of development. In its decision regarding the Kashang Integrated Hydroelectric Project, the NGT directed the government to place the proposed project before the Gram Sabha of the villages before seeking forest clearance for the project. Empowerment of Gram Sabhas by the judiciary validates the powers of the Gram Sabhas under the FRA.

In the Kashang Hydroelectric Project case (*Paryawaran Sanrakshan Sangarsh Samiti Lippa v Union of India*, 2011), the proposed construction exposed the environment as well as the people to risk. The project would not only have impacted the community adversely but also the individual interests of local people. The potential widespread repercussions of the undertaking drew passionate and vociferous resistance from those who would have been directly affected by the project. For instance, the residents of Lippi, a village in Kinnaur District, criticized the Kashang project's environmental impact assessment (EIA) for its failure to account for the relevance of the Pajer and Kerang streams and their role in the survival of the village. The decision of the NGT directing that clearance must be sought from Gram Sabhas endorsed the principle of free and prior informed consent of communities (Dutt, 2016). The application of this principle by the NGT minimizes conflict by seeking and guaranteeing the consent of all parties involved. Moreover, this decision epitomizes both recognitional and procedural justice, recognizing and allowing for the participation of local communities that would be directly affected by the project.

In the case of *The Goa Foundation v Sesa Sterlite Ltd.* (2013), the NGT looked deeper into how mining corporations in India have flouted environmental norms and disproportionately mined forest areas with no compensatory afforestation (required by the law). The case eventually led to the shutting down of 88 mining companies. Tribals, marginal farmers and other oppressed people such as Dalits would benefit most from compulsory afforestation and forest conservation. Many scholarly works have also identified how the tribal peoples live in a symbiotic relationship with the forest, and how any activity that harms this balance could irreparably damage both the forest and people (Bhuskute, 1989).

In contrast to the NGT, the Supreme Court seems to lack clarity in the balance between development and environment. On one hand, in the case of

Narmada Bachao Andolan v Union of India (2000), the Supreme Court seems to have taken a strong pro-development approach. It said that the benefits of the Sardar Sarovar Dam justified the negative effects caused to nearby people and the environment. It further noted that if the dam were not built, the long-term cost to the people would be great, and the absence of a source of income would put increased pressure on the environment. Without the dam, there was no solution to the increasing deprivation of people living in dry areas of Gujarat; with the project, the ratio of benefitted people to adversely affected ones was 100:1. On the other hand, in cases such as *Manchegowda v State of Karnataka* (1984) and *Lingappa Pochanna v State of Maharashtra* (1985), the court has put tribals over and above the development cause adopted by the state, restoring the lands back to tribal peoples even though there had been a transfer of title to a private party. The Supreme Court would profit from a more coherent and structured method to tackle issues of distributive justice in areas of decisional review so as to ensure consistency.

Environmental impact assessment

EIA is a process for predicting and avoiding or mitigating the negative effects of projects with potentially significant environmental and social impacts. It is a mechanism of "dispersed enforcement" (Rosencranz, Janghu and Reddy, 2019), a crucial procedural obligation and a key aspect of due diligence conducted prior to the initiation of a developmental project.

The NGT has considered EIA in numerous cases. For example, it dealt with a challenge to the environment clearance granted by the Union Government to POSCO (*Prafulla Samantra v Union of India*, 2012), a South Korean steel manufacturer. This case involved a Memorandum of Understanding (MoU) for setting up an integrated steel plant, an investment worth approximately US$12 million. The MoU stipulated that the Government of Odisha would expedite POSCO's process for obtaining a clearance from the State Pollution Control Board. The appellants in the case were an environmental activist from Odisha and a farmer who would be directly affected by the proposed undertaking. The appellants challenged the clearance granted by the government, stating that while EIA procedures, including examination by the Expert Appraisal Committee, were seemingly followed, the full letter and spirit of the EIA Notification 2006 were not followed. Hence, while the initial clearance in 2007 was granted, the reappraisal in 2010–2011 found the initial clearance to be lacking. The Government of Odisha at that point had to make a decision on whether or not to regard the new appraisal to be the best authority on the issue. Following the reappraisal, the ministry did not pay heed to the committee that had been set up and validated the clearance.

Consequently, the tribunal found that the clearance granted to POSCO had been obtained by faulty means and suspended the clearance. The NGT stated that the clearance had been granted by an official who was acting in the interests of the corporation and to justify previous administrative decisions he had made. The NGT held that the state government had appointed an official who was making decisions that were coloured by personal biases and that this was a violation of the principles of natural justice. Moreover, the tribunal agreed that projects involving robust foreign investment were imperative to progress and created employment opportunities for the people of India. However, the tribunal went on to state that such developments cannot be undertaken at undue and unnecessary expense of the environment. The NGT decision meant that the South Korean corporation would have to apply for a fresh clearance, unblemished by controversy and bias. Additionally, the ministry would have to establish new terms of reference for the EIA.

Rehabilitation and compensation

India's approach to recognizing Indigenous peoples and their rights has been inconsistent. On the one hand, India voted in favour of the United Nations Declaration on the Rights of Indigenous Peoples as well as the UN International Labour Organization Convention 107 (Indigenous and Tribal Populations Convention); on the other hand, the government continues to deny the concept of "Indigenous peoples," claiming that all Indians are Indigenous (Bijoy, Gopalakrishnan and Khanna, 2010).

Despite having various constitutional and statutory provisions in place to safeguard the rights of Indigenous peoples, such as ratification of 18 human rights treaties, including the Convention on the Elimination of All Forms of Racial Discrimination, harassment and ostracization of tribes and marginalized groups continue to occur, with the portrayal of Indigenous people as backward and downtrodden communities (Cultural Survival, 2016).

With the construction of dams and enormous hydro projects, resettlement of neighbouring tribal people becomes necessary. Resettlement focuses on the physical relocation of the affected group instead of paying attention to their social and economic development. The risks that such people face are seldom a component in the project analysis (Cernea, 1998).

Resettlement seeks to strengthen the position of negatively affected people but is marred by institutional weaknesses, confusion within designated departments and shortfall in capacity. Participation of the affected people is minimal, and they are treated as inconsequential by those overseeing such vast projects. An ideal model of rehabilitation would be marked by the movement from

- landlessness to land-based resettlement;
- joblessness to re-employment;
- food insecurity to safe nutrition;
- homelessness to house reconstruction;
- increased morbidity and mortality to improved health and well-being; and
- social disarticulation and deprivation of common property resources to community reconstruction and social inclusion (Cernea, 1998).

For rehabilitation to happen smoothly, development is a necessary step. Resettlement should ideally be a component of a well-planned development project. Therefore, rehabilitation would be an outcome of resettlement that was formulated not as a mere physical relocation from the impacted area or giving employment to make up for losses to income, but as including development, both social and economic (Jain, 1999).

Specifically, while referring to hydroelectric projects and construction of dams, compensation has been understood as taking monetary measures in the recognition of a loss suffered by persons who have been displaced. More often than not, this comes in the form of a lump sum payment in cash or in kind, largely to benefit the negatively impacted persons (Bartolome, de Wet, Mander and Nagraj, 2009).

There are broad-ranging concerns in respect of awarding compensation to such people. The losses incurred by such tribal and other Indigenous people are not properly accounted for; hence there is no fair compensation (Bartolome, de Wet, Mander and Nagraj, 2009). The people living downstream are the ones who are most severely affected since their lands are taken over for compensatory afforestation. Furthermore, compensation is awarded only to those people who possess an undisputed legal title. Paradoxically, tribal people, labourers and encroachers are hardly considered eligible for compensation, even when they are the most vulnerable. Ultimately, the limited provisions of law regarding challenges to the rate of compensation are inaccessible to such people, since they are unaware of the legal nuances and cannot afford the expensive process of adjudication.

The major drawback of compensation is that it solely addresses the loss of material things such as property and assets but fails to consider rights (Bartolome, de Wet, Mander and Nagraj, 2009). Compensation on the basis of replacement value is still restricted to individually owned property; the totality of rights that are violated is not compensated. For example, the Fact-Finding Committee on the Srisailam Project found that the replacement value of one acre of dry land was around Rs. 5000, and for one acre of wetland, Rs. 13,800. The compensation actually paid was only Rs. 932 and Rs. 2332, respectively. Thus, the amount paid as compensation was merely one-fifth the amount that would be needed by the negatively affected persons to purchase agricultural land of equivalent quantity as well as quality.

The NGT has made every effort to recognize the rights of poor tribal people and emphasized the need to allow development within the ambit of environmental regulations. In the case of *Lower Painganga Dharan Virodhi v State of Maharashtra* (2014), the NGT was tasked with adjudicating a dispute relating to the approval of a dam project that would have immense economic, social and environmental repercussions. The applicants, a group of tribal farmers and villagers, approached the tribunal to quash the environmental clearance issued to the project, claiming that it was obtained illegally: the procedural requirement of affording a public hearing had not been followed.

After perusing a plethora of documents and records, the tribunal observed that while irrigation facilities in the area would assist local cultivators by reducing their dependency on rainfall, the community's right to development would also include the protection of fundamental human rights. Therefore, if the objective of the irrigation was to benefit the community, it must ensure that the "Project Affected Families" are given a rehabilitation package which would not only provide basic amenities such as food and shelter but also restore the community's means of livelihood (*Lower Painganga Dharan Virodhi v State of Maharashtra*, 2014). Furthermore, the Comptroller and Auditor General's report on the Management of Irrigation Projects in Maharashtra condemned the continuous failure on part of the state government to undertake environmental protection (Dandekar, 2014).

It is clear that to ensure that restorative justice is served, the NGT must be careful when calculating compensation, taking into account factors that may not be fully captured by the law that exists and must cast a duty upon the government to be as proactive as possible in compensating and rehabilitating individuals at risk.

Setting up committees

To ensure the veracity of the claims, gather more information and properly implement its orders, the NGT has set up committees and panels of experts (see for instance *Manoj Mishra v Union of India & Others*, 2018). There is no express power given to the NGT under the NGT Act to constitute such committees. Such power has been sanctioned by the Supreme Court under Section 3(3) of the Environment (Protection) Act 1986 (*S. Jagannath v Union of India*, 1997). However, a few scholars have criticized the move, as they believe that setting up such committees makes the NGT "investigative" rather than "adjudicative" (Rosencranz and Sahu, 2014). As the tribunal is a part of the judiciary, its role must be restricted to the latter.

In 2018, the NGT considered the pollution of the Son River in *Nityanand Mishra v State of M.P. & Others* (2018). The illegal sand mining activity along the banks was found to be critically affecting the flow of the river. The case was only brought to light because of the extinction-level threat to gharials (fish-eating crocodiles). The tribunal immediately issued directions to stop this illegal pollution for the protection of the river and the wildlife near the Bansagar Dam and constituted a committee to oversee the directions of the tribunal. The NGT in another 2018 case created a task force that would ensure that no illegal sand mining would take place in riverbeds of polluted stretches of land (*Indian Council for Enviro-Legal Action v National Ganga River Basin Authority & Others* and *M.C. Mehta v Union of India & Others*, 2015).

In the case of Threat to Life Arising Out of Coal Mining in South Garo Hills District v State of Meghalaya & Others (2018) the tribunal decided the issue of rat hole mining, a dangerous and harmful practice. This form of mining is known as such because of the rather small holes that are dug up to let the mining happen. Not only is this practice extremely harmful to the environment, but also it has none of the safeguards afforded to the miners. The tribunal established a committee, the interim report of which found that mining activity in the area had been going on for a long time in an unplanned and unscientific manner, creating grave ecological harm. To improve compliance in the industry, the NGT imposed an interim penalty of Rs. 100 crores on the Government of Meghalaya. In doing so, the tribunal established the precedent that a state can be made liable for colluding with polluters as well as being charged with non-compliance, which in itself is a victory for environmental justice (Gill, 2016). To uphold distributive and procedural justice, it is crucial that these committees are properly representative of the various interests at stake, and that participation is ensured for parties who are not able to afford coming to the court. Anuj Bhuwania has chronicled the troubling aspects of how these committees operate in his seminal work *Courting the People* and argued that committees often jettison the interests of the people who are most directly and acutely affected by the scheme due to their lack of representation in these committees (Bhuwania, 2016).

Fisherfolk

According to the Central Marine Fisheries Research Institute Census 2010, the total marine fisherfolk population in India was about 4 million, comprising 864,550 families (Department of Animal Husbandry, Dairying and Fisheries, 2010). A whopping 61 per cent of these families were in the Below Poverty Line category. Given the age of these data, the number of fisherfolk is probably well above the already high 4 million. This population primarily lives and operates on the ecologically sensitive coast. About 63.6 per cent of the fisherfolk were engaged in fishing and allied activities. Fisherfolk are impacted by various industrial and commercial activities on the coast such as the construction of ports, thermal power plants, Special Economic Zones, resorts, hotels and real estate projects.

Background to fisherfolk's rights and legislation in India

Much fishery-related legislation exists in India at both the national and state level. Most coastal states have formulated the Marine Fishing Regulation Acts after a model bill was circulated by the Government of India. Some of the early adopters were Kerala and Goa in 1980, while Gujarat, Andaman and Nicobar Islands and Lakshadweep enacted the legislation as late as 2004 (Datta, 2013). The other states passed legislation in the period between these two extremes. These acts demarcate areas in territorial waters for mechanized and non-mechanized vessels, as well as specify what boats and nets can be used in different areas.

To acknowledge the stake that fisherfolk have in the coast, the Coastal Regulation Zone Notification 2011 was brought in. Its stated objective was to ensure the livelihood security of fisher and other rural communities in coastal areas while conserving and protecting coastal stretches (Ministry of Environment and Forests, 2011). This notification ensured that activities involving the coast were monitored by bodies that included representatives from fishing communities, among other provisions that ensured the involvement of fisherfolk. It also permitted fisherfolk to live in more sensitive areas of the coast as long as there were comprehensive plans for sanitation and disaster management.

Response of the NGT to the fisherfolk's cause

Given that the NGT came into existence in 2010, a major chunk of the litigation involving fisherfolk relates to the construction in violation of coastal zone regulations. These cases often involve challenging the approval for construction granted by the Coastal Zone Management Authority. A variant of these cases is where the procedure followed is in doubt, when stakeholders have not been properly consulted while an impact assessment was carried out (*Ossie Fernandes v Ministry of Environment & Forest*, 2011). Another species of litigation is when fisherfolk claim compensation for direct damage to their ecosystem by negligent activities that result in pollution, such as an oil spill (*Samir Mehta v Union of India*, 2011) or a discharge of chemical effluents (*Kasala Malla Reddy and Others v State of Andhra Pradesh and Others*, 2013).

Often to gain legitimacy as representatives of fisherfolk, the petitioners will try to establish membership in a union or welfare board. An instance of this is *Wilfred J. v Ministry of Environment & Forests* (2014), where the petitioners demonstrated that they were members of the Fish Workers Welfare Board formed by the Government of Kerala. The NGT has to be wary of different parties claiming to represent the interests of fisherfolk communities. In *Octavia Albuquerque and Others v Union of India* (2013), the petitioners tried to argue that the Mangalore Fishery Harbour project would lead to the filling up of a creek, to the detriment of fisherfolk repairing their boats there. However, the court noted that, in actuality, the local fisherfolk wanted the execution of the project. They got their association impleaded in the appeal, contending that the project would give more facilities for the fisherfolk, both for parking and repairing the vessels.

Another type of misrepresentation is by the government, claiming that its actions are for the good of the fisherfolk. This was seen in the case of Udaya Suvarna and Others v The Deputy Commissioner (2016), where the government had granted permission for the removal of sandbars from rivers in Udupi District in Karnataka (see Chapter 11, this volume), purportedly because the sandbars were an obstruction to navigation and fishing. The court observed that the removal of the sandbars was not in the interest of the fisherfolk but merely a facade to justify illegal sand mining activities. The court has been awake to the danger of the interests of fisherfolk being misrepresented by various parties and it must continue to be so, to ensure that procedural and recognitional justice is achieved.

The NGT has also taken a pragmatic approach to distributive justice, striking a fine balance between protecting the environment and ensuring that development is not unduly hampered. In *Wilfred J. v Ministry of Environment & Forests* (2014), the tribunal emphasized that even if a project was not financially viable and would lose money in the foreseeable future, it could still have larger economic benefits. For instance, though the actual recovery of investment from the running of a metro rail in a metropolis may be insignificant, the expansion of the economy as a consequence of residents being able to travel around safely and rapidly is what makes the project feasible in a full cost–benefit analysis. These economic benefits need to be balanced with environmental costs in ascertaining the justifiability of a project.

When it comes to compensation, the tribunal has in specific instances awarded large sums. Generally, penalties imposed by the tribunal are directed to be paid to the affected communities or the concerned authorities to invest in remedying the damage caused to the environment. However, these sums are often not commensurate with the damage caused. The two landmark cases in which the court has awarded compensation or levied fines for affecting the livelihood of communities are Ramdas Janardan Koli v Ministry of Environment and Forests (2013) and Hazira Macchimar Samiti v Union of India (2013).

In *Ramdas Janardan Koli*, the NGT observed that certain port expansion activities had affected the livelihood of the fishing communities of the Raigad District. Some 1630 families had been affected. The tribunal determined the compensation to be awarded to be Rs. 95 crores. Though this amount seems considerable, the calculations provided for just three years' worth of compensation estimated at only Rs. 532 per day per family. This per diem amount is worthy of criticism, as it is less than the minimum wage stipulated under the Mahatma Gandhi National Rural Employment Guarantee Act. At no point in the calculation does the tribunal refer to any external source to anchor the figures that they have assumed in arriving at the compensation. If the compensation is the sole tangible result after such a long judgement, for the sake of restorative justice, care must be exercised that the value is a robust one, taking into account as much data as possible.

In *Hazira Macchimar Samiti v Union of India* (2013), the tribunal imposed a fine of Rs. 25 crores on Adani Hazira Port Private Ltd. for ecological damage and the impact on fishing communities. The tribunal in the past has made it clear that the polluter pays principle extends not only in compensating the victims of pollution but also the cost of remedying the environmental degradation (*Indian Council for Enviro-legal Action v Union of India*, 2011). However, the tribunal does not seem to have included any calculations explaining the fine issued in this case, let alone any division of the costs between the damage to the environment and the communities. The penalty also seemed to be a paltry amount in comparison to the total project cost of Rs. 1800 crores.

Policy reform is needed to assist the tribunal in calculating accurate compensation amounts. A comparative study of ecological compensation programs in the United States, England, Germany and South Africa has suggested two design aspects that have a considerable influence on the outcomes of these programmes; the integration of compensation amounts with conservation schemes themselves and an open-access centralized reporting system for the collection of data on various projects around the country (Koh, Hahn and Ituarte-Lima, 2014). These systems, coupled with judicial precedent, can create a robust method that will hopefully one day be as developed as techniques used in land acquisition cases. This would serve to advance restorative justice in India.

Recently, the Government of India floated the Draft Environment Laws (Amendment) Bill 2015 (MoEFCC (Ministry of Environment, Forest, and Climate Change), 2015), which proposed to amend the Environment (Protection) Act 1986 (EP Act) and the NGT Act. To rationalize the process of determining compensation, the bill seeks to classify the damages into "minor violations," "non-substantial damage" and "substantial damage" on the basis of the severity of the incident. The bill has inserted minimum and maximum levels of penalty that can be levied in each of these cases (Environment (Protection) Act, 1986, 3–5). It further seeks to set up an "adjudicatory authority" under Section 3(3) of the EP Act for such incidents. This authority would take into consideration factors such as the amount of damage to the environment, the disproportionate gain or unfair advantage earned and the injury caused to people, property and living creatures. The NGT has been conferred appellate jurisdiction over the determination of the penalty. Although the constitution of a separate body and laying down of guidelines will reduce the exercise of arbitrary discretion, there are certain inherent problems with the bill (see for instance Trivedi, 2015).

The tribunal must take the utmost care to ensure that the interests of fishing communities are truly represented by the parties before them. Often fishing communities cannot afford the same level of representation as other parties to the petition and may find that they are being ignored in arguments before the court. The tribunal must also try to be as transparent as possible in its awarding of compensation. It is not enough to award a large amount; the amount must be cogently determined. With the lax coastal zone regulations approved by the central government cabinet in December 2018 (Government of India, 2018), the tribunal must take care that the interests of fisherfolk are not trodden over in the pursuit of development.

Conclusion

Although the judiciary has frequently attached greater importance to economic and technological progress, the NGT has proved itself to be a people's court. The forum has not entirely sidelined economic advancement, even while striving to protect environmental interests and resources. Sceptics might question the real tangible efficacy of the NGT, but the fortitude and maturity of the forum is without doubt admirable. The NGT has carved for itself an image of being an axis in the application of sustainable development. As an institution, it postulates a pragmatic development model that embraces maximum human development while substantially minimizing irreversible damage to the environment. The NGT must continue to do so while ensuring that its decisions are reasoned coherently and consistently, so as to guarantee the delivery of recognitional, procedural, distributive and restorative justice.

References

Amnesty International. (2016) *"When Land Is Lost, Do We Eat Coal?" Coal Mining and Violations of Adivasi Rights in India.* Bangalore: Amnesty International India. Available at: https://www.amnestyusa.org/wp-content/uploads/2017/04/coal_report_final.pdf.

Bartolome, L.J., de Wet, C., Mander, H. and Nagraj, V. (2009) *Displacement, Resettlement, Rehabilitation, Reparation and Development.* World Commission on Dams Thematic Review 1.3. Available at: http://citeseerx.ist.psu.edu/viewdoc/download?-doi=10.1.1.112.9372&rep=rep1&type=pdf (Accessed: 14 December 2020).

Bhuskute, R.V. (1989) "Tribals, Dalits and government lands," *Economic and Political Weekly*, 24(42), 2355–2358.

Bhuwania, A. (2016). *Courting the People: Public Interest Litigation in Post-Emergency India.* Cambridge: Cambridge University Press.

Bijoy, C.R., Gopalakrishnan, S. and Khanna, S. (2010) *India and the Rights of Indigenous Peoples: Constitutional, Legislative, and Administrative Provisions Concerning Indigenous and Tribal Peoples in India and Their Relation to International Law on Indigenous Peoples.* Asia Indigenous Peoples Pact.

Bose, I. (2010) *How Did the Indian Forest Rights Act, 2006, Emerge? Improving Institutions for Pro-Poor Growth.* IPPG Discussion Paper Series. Available at: https://www.gov.uk/research-for-development-outputs/how-did-the-indian-forest-rights-act-2006-emerge (Accessed: 14 December 2020).

Government of India. (2018) "Cabinet approves coastal regulation zone (CRZ) notification 2018," Press Information Bureau, 28 December. Available at: http://pib.nic.in/newsite/PrintRelease.aspx?relid=186875.

Cernea, M.M. (1998) "Impoverishment or social justice? A model for planning resettlement" in Mathur, H.M., Marsden, D. (eds.) *Development Projects and Impoverishment Risks.* New Delhi: Oxford University Press, 42–66.

Council for Social Development. (2010) *Summary Report on Implementation of The Forest Rights Act.* Available at: http://www.indiaenvironmentportal.org.in/files/Council_for_Social_Development_final_summary_report.pdf (Accessed: 14 December 2020).

Cultural Survival. (2016) "Observations on the state of indigenous human rights in India." Prepared for The United Nations Human Rights Council Universal Periodic Review 2016 (27th Session), Third Cycle. Cambridge, MA: Cultural Survival. Available at: https://www.culturalsurvival.org/sites/default/files/INDIAUPR2016final.pdf (Accessed: 14 December 2020).

Dandekar, P. (2014) "Rampant environmental violation of Maharashtra Water Resource Department," press release, South Asia Network on Dams, Rivers and People, 21 June. Available at: https://sandrp.in/2014/06/20/press-release21-06-14-

rampant-environmental-violations-of-maharashtra-water-resource-department (Accessed: 9 May 2019).

Datta, D. (2013) *Fisheries Legislation in India*. Central Institute of Fisheries Education, Kolkata. Available at: https://www.researchgate.net/publication/270394242_Fisheries_Legislation_in_India (Accessed: 14 December 2020).

Department of Animal Husbandry, Dairying and Fisheries (2010) *Marine Fisheries Census 2010*. New Delhi: Ministry of Agriculture, Government of India. Available at http://eprints.cmfri.org.in/8998/1/India_report_full.pdf (Accessed: 15 February 2012).

Dutt, B. (2016) "Will empowering Gram Sabhas derail development?" *Mint*, 13 May. Available at: https://www.livemint.com/Opinion/wttwdia1Uxnf9h3QJ2fVzO/Will-empowering-gram-sabhas-derail-development.html (Accessed: 12 April 2019).

Environment (Protection) Act 1986, No. 29, Acts of Parliament, 1986 (India).

Forest (Conservation) Act 1980, No. 69, Acts of Parliament, 1980 (India).

Gill, G.N. (2016). *Environmental Justice in India: The National Green Tribunal*. Abingdon: Routledge.

Hazira Macchimar Samiti v Union of India, Appeal No. 79/2013 (National Green Tribunal).

Indian Council for Enviro-Legal Action v National Ganga River Basin Authority & Others and M.C. Mehta v Union of India & Others, O.A. No. 10/2015 (National Green Tribunal).

Indian Council for Enviro-legal Action v Union of India (2011) 8 SCC 161 (India).

Jain, L.C. (1999) *Comments on Draft Thematic Review*, Sao Paulo Minutes of the Commission Meeting (World Commission on Dams), September 1999.

Kasala Malla Reddy and Others v State of Andhra Pradesh and Others, Application Nos. 69, 70, 71, 72, 86, 87, 88, 89, 91, 82 and 90/2013 (SZ) (THC) (National Green Tribunal).

Koh, N.S., Hahn, T. and Ituarte-Lima, C. (2014) *A Comparative Analysis of Ecological Compensation Programs: The Effect of Program Design on the Social and Ecological Outcomes*. Working Paper, Uppsala University. Available at: https://uu.diva-portal.org/smash/get/diva2:772933/FULLTEXT01.pdf.

Lingappa Pochanna v State of Maharashtra, AIR 1985 SC 389 (Supreme Court of India).

Lower Painganga Dharan Virodhi v State of Maharashtra, Appeal No. 13/2013 (2014) (National Green Tribunal WZ).

Manchegowda v State of Karnataka, AIR 1984 SC 1151 (Supreme Court of India).

Manoj Mishra v Union of India & Others, Original Application No. 06 of 2012 (2018) (National Green Tribunal).

Ministry of Environment and Forests (2011) Coastal Regulation Zone Notification 2011, 6 January. Available at: http://moef.gov.in/wp-content/uploads/2017/08/7_0.pdf.

MoEFCC (Ministry of Environment, Forest, and Climate Change) (2015) Draft Environment Laws (Amendment) Bill, 2015, 7 October. Available at: http://www.indiaenvironmentportal.org.in/files/file/Draft%20Environment%20Law.pdf.

Narmada Bachao Andolan v Union of India (2000) 10 SCC 664 (Supreme Court of India).

Nityanand Mishra v State of M.P. & Others, O.A. No. 456/2018 (2018) (National Green Tribunal). Available at: http://www.greentribunal.gov.in/Writereaddata/Downloads/456-2018(PB-I)OA31-7-18.pdf (Accessed: 9 May 2019).

Octavia Albuquerque and Others v Union of India, Department of Animal Husbandry, Dairying and Fisheries, Ministry of Agriculture and Others, Appeal No. 25/2013 (THC) and M.A. No. 166/2013 (National Green Tribunal).

Ossie Fernandes v Ministry of Environment & Forest, Appeal No. 12/2011 (National Green Tribunal).

Paryawaran Sanrakshan Sangarsh Samiti Lippa v Union of India, Appeal No. 28 of 2013 (2011) (National Green Tribunal).

Praffulla Samantra v Union of India, Appeal No. 8 of 2011 (2012) (National Green Tribunal).

Ramachandran, N. (2016) "One institution at a time," *Mint*, 13 June. Available at: https://www.livemint.com/Opinion/yMZClKPH1uVvpwIUxi7DgL/One-institution-at-a-time.html (Accessed: 13 August 2019).

Ramdas Janardan Koli and Others v Secretary, Ministry of Environment and Forests and Others, Application No. 19/2013 (National Green Tribunal).

Rosencranz, A. and Sahu, G. (2014) "Assessing the National Green Tribunal after four years," *Journal of Indian Law and Society*, 5(Monsoon), 191–200.

Rosencranz, A., Janghu, S. and Reddy, P. (2019) "The evolution and influence of international environmental norms," *Environmental Law Reporter*, 49(2), 10133.

S. Jagannath v Union of India, AIR 1997 SC 811 (Supreme Court of India).

Samir Mehta v Union of India, Original Application No. 24 of 2011.

Sarin, M., Singh, N.M., Sundar, N. and Bhogal, R.K. (2003) *Devolution as a Threat to Democratic Decision Making in Forestry? Findings from Three States in India.* Working Paper 197, Overseas Development Institute. Available at: https://www.odi.org/sites/odi.org.uk/files/odi-assets/publications-opinion-files/2436.pdf.

Sarin, M. and Springate-Baginski, O. (2010) *India's Forest Rights Act – The Anatomy of a Necessary but Not Sufficient Institutional Reform.* IPPG Discussion Paper Series Number Forty-Five. Available at: https://assets.publishing.service.gov.uk/media/57a08b-0be5274a27b2000909/dp45.pdf (Accessed: 14 December 2020).

Springate-Baginski, O., Sarin, M., Ghosh, S., Dasgupta, P., Bose, I., Banerjee, A., Sarap, K., Misra, P., Behera, S., Reddy, M.G. and Rao, P.T. (2009) *Redressing "Historical Injustice" through the Indian Forest Rights Act 2006: A Historical Institutional Analysis of Contemporary Forest Rights Reform, Improving Institutions for Pro-Poor Growth.* IPPG Discussion Paper Series Number Twenty-Seven. Available at: https://assets.publishing.service.gov.uk/media/57a08b66e5274a27b2000b05/dp27.pdf (Accessed: 14 December 2020).

The Goa Foundation v Sesa Sterlite Ltd., M.A. No. 49 of 2013 (National Green Tribunal).

The Scheduled Tribes and Other Traditional Forest Dwellers (Recognition of Forest Rights) Act 2006, No. 2, Acts of Parliament, 2007 (India).

The Wildlife (Protection) Act 1972, No. 53, Acts of Parliament, 1972 (India).

Threat to Life Arising Out of Coal Mining in South Garo Hills District v State of Meghalaya & Ors, O.A. No. 673/2018 (National Green Tribunal).

Trivedi, D. (2015). "What are the proposed changes under Environment Laws (Amendment) Bill, 2015?" *Pleaders: Intelligent Legal Solutions*, 7 November. Available at: https://blog.ipleaders.in/changes-under-environment-laws-amendment-bill-2015/.

Udaya Suvarna and Others v The Deputy Commissioner/Chairman District Sand Monitoring Committee and Others, Application No. 111 of 2016 (SZ) and M.A. Nos. 133, 136, 138/2016 (National Green Tribunal).

Wilfred J. v Ministry of Environment & Forests, M.A. No. 182/2014 & M.A. No. 239/2014 in Appeal No. 14/2014 and M.A. No. 277/2014 in Original Application No. 74/2014 (National Green Tribunal).

6   Being appraised by experts

A review of the role of Expert Appraisal Committees in the environmental clearance process and judicial intervention

Himanshu Pabreja and Neelotpalam Tiwari

Introduction

The Bhopal gas disaster[1] was a watershed moment in Indian environmental jurisprudence, for it highlighted the ongoing incapacity of Indian authorities in dealing with environmental disasters. To some extent, on account of the shambolic response of authorities, it became a tipping point that prompted the entry of the global environmental justice movement into India, demanding stronger environmental laws capable of safeguarding every section of society from environmental hazards. Environmental justice intends to secure the right to a safe, healthy and productive environment for all (Yang, 2002). In addition, it covers principles for securing these rights, namely, participation of different stakeholders in environmental governance (Schlosberg, 1999, 23–25); equitable distribution of risks (Foster, 2002, 461), benefits and costs of environmental decision-making (Cole and Foster, 2001, 16–20); rights of the public to access justice; and restoration following adverse impacts of activities on the environment (Jenkins, 2018, 51–52). As the gas leakage tragedy showcased the extent to which Indian legislation [the Air (Prevention and Control of Pollution) Act 1981; the Water (Prevention and Control of Pollution) Act 1974; and the Forest (Conservation) Act 1980] fell short on these aspects, demand arose for a comprehensive legislative structure that would cater to emerging environmental issues as well as future development needs while at the same time complying with basic tenets of environmental justice.

Subsequently, the Indian government responded with the enactment of the Environment (Protection) Act 1986 (EPA), vesting with the central government the power to formulate rules and regulations to deal with environmental issues as well as determine environmental standards in India. Under the EPA, government promulgated the Environment Impact Assessment (EIA) Notification in 1994, which was later superseded by the EIA Notification 2006[2] (S.O. 1533, 2006) (EIA Notification), for the purpose of assessing probable environmental impacts of a project before its commencement. While

DOI: 10.4324/9781003141228-6

in theory, the EIA Notification gave statutory recognition to demands for environmental justice by the public, in practice, its success in achieving its objectives has been contentious.

The EIA Notification created a four-step procedure, consisting of screening, scoping, public consultation and appraisal. Screening provides for the categorization of projects (Category A, Category B1, Category B2) based on their size (capacity), nature and location, which determine whether EIA is required and if so, the type of assessment that should be done. This is followed by scoping, which determines the range of considerations and the types or scope of studies that should be part of the EIA for any given project. Then, public consultation aims to ensure representation of the public in environmental decision-making, and appraisal involves expert-level consideration of multiple concerns related to the project. Lastly, a final decision is taken by the Ministry of Environment, Forest and Climate Change (MoEFCC) or State Environment Impact Assessment Authority (collectively referred as "Regulatory Authorities"), as the case may be, based on the expert-level consideration of the project.

These steps signify that the EIA Notification aims to ensure thorough consideration of the project to identify potential environmental concerns that may arise due to its commencement; disclosure of the correct and relevant information about the project to the public, and an invitation to the public to express their concerns on it; and assessment of such information and concerns related to the project by a multidisciplinary expert committee for final decision-making. Keeping these aspects in mind, it is also important to highlight that a fundamental objective of law is to ensure justice for stakeholders involved in the process. The same also stands true for the environmental law: its relevance for society ought to be assessed on the extent to which it is consistent with fundamental principles of environmental justice, such as procedural, recognitional, distributive and restorative justice. Accordingly, an assessment of the EIA Notification on this basis assumes relevance for this chapter.

A well-defined multistep procedure in the EIA Notification not only provides procedural certainty and transparency to the EIA process but also ensures the involvement of different stakeholders in the same. As such, it seeks to serve the objective of procedural justice by providing the public an opportunity to participate by raising their concerns about a project as well as demanding serious appraisal and consideration of their concerns by the authorities. Besides, consideration of these issues by multidisciplinary expert committees under the EIA Notification ensures that procedural involvement of stakeholders can bring certain substantive impact on the outcome. For instance, such consideration by authorities may facilitate distributive justice for different stakeholders in the form of just or equitable distribution among them of environmental risks and benefits.

Furthermore, recognition of the right of stakeholders to approach appropriate judicial authorities to "restore" the damaged situation in case of non-compliance with the EIA Notification opens the way to restorative justice to stakeholders. Additionally, because the scope of the EIA Notification extends even to post-clearance monitoring by authorities and reporting by project proponents about compliance with conditions under which clearance was granted, the notification also prevents the distortion of environmental justice factors the authorities had in mind while granting clearance to a project.

Accordingly, theoretically, each step of the EIA Notification contributes towards securing goals of environmental justice. However, its success in achieving these goals would need to be determined based on the analysis of functioning of authorities responsible for its implementation. In light of their significance in the EIA process, this chapter first critically analyses the role and composition of the authorities defined under the EIA Notification. Second, based on readings of judicial and quasi-judicial pronouncements from 2010 to 2019 (until April 2019), we analyse the grounds for judicial intervention in the environmental clearance (EC) decisions of authorities. Finally, we conclude by providing certain observations and recommendations based on judicial responses to these decisions.

Role and composition of expert committees in the EIA process

The EIA Notification has established the multidisciplinary, sector-specific Expert Appraisal Committees at union (EAC) and state levels (SEAC) (collectively referred to as the "Committees"), which have effectively "outsourced" the duty to govern and administer the EIA procedure (*Utkarsh Mandal v Union of India*, 2009). The EIA Notification mandates the Committees to consider and pragmatically review documents, reports, recordings and other technical and legal considerations [S.O. 1533, 2006, para 7(i)(IV)] related to a proposal and furnish reasoned recommendations to the Regulatory Authorities after due application of mind (*Lower Painganga Dharan Virodhi Sangharsha Samiti v State of Maharashtra*, 2013). These recommendations become crucial as, being sourced from an expert committee, they are often relied upon by the Regulatory Authorities for the final approval or rejection of a proposal. Accordingly, the Committees are expert bodies that form the central point of the entire procedure of the EIA Notification, being involved in all its steps (S.O. 1533, 2006, para 8; Banerjee, 2015).

Given their crucial role in the EIA Notification, it becomes important to ensure that Committee members are fully competent to ponder techno-legal aspects of a project. Therefore, to assess the efficacy of the EIA Notification, we will comment on the composition of these Committees as well as analyse an important dispute concerning eligibility criteria for their members and chairpersons.

Composition of the Committees

The EIA Notification contains detailed provisions about the constitution of the Committees. Para 5 provides for their constitution, and Appendix VI provides eligibility criteria for their members and chairpersons. The Union Government is required to reconstitute the Committees after every three years. The Committees have identical compositions, made up of only "professionals" and "experts" according to the criteria in Appendix VI, subject to a maximum of 15 members, including the chairperson (S.O. 1533, 2006, para 5 read with Appendix VI).

As for the eligibility of the chairperson, Appendix VI stipulates that he or she shall be an "outstanding and experienced environmental policy expert or expert in management or public administration with wide experience in the relevant development sector." Other members shall be experts in fields specified in the EIA Notification. Moreover, the members and chairperson have a limited tenure of two terms of three years each, which provides the opportunity for varied representation on the Committees by ensuring that no person continues on them for an extended period. Also, the chairperson and members are prevented from removal or ouster before the expiry of their term without proper enquiry and specific reasons, to ensure continuity in their service and functional autonomy in their decision-making (S.O. 1533, 2006, Appendix VI).

A conjoint reading of these provisions ensures that Committees have a multidisciplinary character. Since appraisal of a project requires considering multidisciplinary aspects, Committees scrutinizing potential environmental impacts must comprise members from different fields, namely, science, management, law and administration (*Goa Foundation v Goa State Environment Impact Assessment Authority*, 2015). Their multidisciplinary membership ensures their capacity to scrutinize a project from technical, scientific, legal and other relevant perspectives. Given their significance in the process, we next discuss judicial analysis of the Committee composition.

Dispute around committee chairpersons and members

The composition of the Committees was recently subjected to judicial scrutiny by the National Green Tribunal (NGT),[3] which commented on how their makeup influences the effective implementation of environmental policies in India (*Kalpavriksh v Union of India*, 2013). This petition challenged the amendments made in Appendix VI to the EIA Notification by virtue of the notification dated 11 October 2007 (Ghosh, 2013, 438; S.O. 1737 (E), 2007). The NGT had to deliberate (1) about its jurisdiction to adjudicate on this issue and (2) on appropriate directions to the MoEFCC with respect to eligibility criteria for the appointment of Committee members and chairpersons.

On the first aspect, the NGT upheld its jurisdiction under the NGT Act 2010, observing that these Committees perform the most important functions under the EIA Notification, and, thus, their faulty decisions would adversely impact the environment in general. Therefore, on account of their integral role in the EIA Notification, a matter related to the appointment of an ineligible person as a member or chairperson would be a "substantial question related to environment," which the tribunal has jurisdiction to settle under the NGT Act.

On the second aspect, the NGT commented on two significant modifications introduced in Appendix VI by the notification of 11 October 2007. One was the addition of a new criterion, experience in "public administration or management," for appointment to Committees [S.O. 1737 (E), 2007, para IV(i)]. The other was the removal of the only eligibility criteria for appointment as a chairperson of a Committee, as was originally provided in the 2006 Notification [S.O. 1737 (E), 2007, para IV(ii)].

Taking critical note of these modifications, the NGT observed that these Committees are responsible for effective implementation of the EIA Notification by assessing likely impacts of a proposed project on the environment in general. Given that such projects involve complex issues not only concerning the environment but also about rehabilitation, resettlement and so on, it is pertinent that Committee members have the required expertise to assess social and environmental considerations of a project.

Since dilution of their eligibility criteria could result in the entry of people who lack the requisite professional experience, the tribunal observed that appointing persons with only general administration or management experience would be inconsistent with the objectives of the EIA Notification and the EPA. The MoEFCC was accordingly directed to appoint only those persons as members or chairperson who have "experience or expertise" in the environmental sector and specify appropriate eligibility criteria for chairpersons in the EIA Notification.

Subsequently, as directed by the tribunal, Appendix VI of the EIA Notification was amended to incorporate "public administration or management covering various developmental sectors and environmental issues" as an eligibility criterion for a member. Whereas, a requirement of "experience in environmental policy related issues, in management or in public administration dealing with various developmental sectors" was added for the chairperson (S.O. 1533, 2015, Appendix VI, p. 51, clause 2 and 4).

An interesting aspect to note is that, although the NGT's order resulted in these rectifications in the EIA Notification, in practice, the inclination of the government to appoint persons with managerial skills rather than environmental expertise continued to be problematic (Banerjee, 2015). Moreover, despite making critical comments against dilution of eligibility, the NGT did not consider it appropriate to set aside appointments already made to Committees and decisions taken by them from the date of the modifications until its judgement.

Critical role in the EIA process: data on functioning of the EAC from 1986 to 2009

A study of the functioning of the Committees under the EIA Notifications 1994 and 2006 is relevant for understanding their crucial role in the EIA process. As such, a perusal of selected findings with respect to the number of projects considered by the EAC assumes significance (Menon and Kohli, 2009, 20):

- The number of projects that were granted clearance by the MoEFCC from 1986 to 2006, based on the original Notification, was **4016**, which comes to around **192** projects each year.
- However, this number escalated exponentially from September 2006 to August 2008, as the MoEFCC granted clearance to **2019** projects, close to **1010** projects each year. Moreover, the MoEFCC granted clearance to a staggering **2586** projects between January 2008 and February 2009.
- Clearing close to 2600 projects in a year is indeed a tedious task. However, the amount of time spent on each project by the EAC was around **12** minutes per project.
- From September 2006 to August 2008, the rejection rate of the projects by the EAC was close to **1.2** per cent. This number includes any projects that were sent back for reconsideration.
- The situation did not improve, as the project rejection rate dropped further to **0.8** per cent between January 2008 and February 2009.

These numbers support an inference that the critical functioning of the EAC is being adversely affected by it being overworked, through having long lists of proposals to be considered in only a limited number of meetings. This severely affects the quality of deliberations concerning a project. Furthermore, it also implies a reduction of quality in the appraisal procedure, as evident from the reduced time given to each project and the low rate of rejections.

In light of this grim picture about the number of projects considered by the EAC, we next analyse the broader function of the Committees in the EIA process as well as judicial and quasi-judicial decisions (from 2010 to April 2019), reviewing their role and function.

Appraisal by the Committees

Generally speaking, "appraisal" means an exercise to estimate the fair price, value or worth of a proposed activity (Garner, 2019). Under the EIA Notification, appraisal signifies a "detailed scrutiny" by a Committee of the information provided by project proponents. This stage requires proponents to submit the "[f]inal EIA report, a copy of a video tape of the public hearing, a copy of the final layout plan, and a copy of the project feasibility report" to the appropriate Regulatory Authority for final appraisal by Committees, which shall be completed within 60 days of receipt of the relevant documents [S.O. 1533, 2006, para 5, 7(i)(IV) and Appendix V].

In terms of the EIA Notification, the Committees are required to pragmatically consider all relevant materials related to a proposal; conduct a detailed review of these documents; review outcomes of the public consultation process; strictly scrutinize the final EIA report and other documents with reference to the terms of reference issued at the initial stage to identify any inconsistencies or absence of material information (S.O. 1533, 2006, Appendix V, point 2); honour the proponent's right to furnish clarifications to queries raised about a project before taking the final decision (*Lower Painganga Dharan Virodhi Sangharsha Samiti v State of Maharashtra*, 2013); specify safeguards or conditions on an approved project; and finalize and release minutes of its meetings. In order to ensure that the appraisal step does not become a mere formality, the EIA Notification emphasizes that Committees must furnish their reasoned recommendations (S.O. 1533, 2006, Appendix V, point 6) on a proposal because as the only expert advisory body in the EIA procedure, their reasons are given significant weight by the Regulatory Authorities in their final decisions [S.O. 1533, 2006, para 8(ii)]. Cumulatively, these aspects attempt to secure transparency and accountability in decision-making by the Committees (*Goa Foundation v Goa State Environment Impact Assessment Authority*, 2015).

That said, the EIA Notification has placed a special emphasis on the Committees as a "weighing scale" for balancing ecological integrity and the development needs of a region. It is a professional entity performing a delegated task of evaluating relevant factors related to a project. Based on this evaluation and keeping the precautionary principle in mind, the Regulatory Authority may attach any number of safeguards or conditions to the EC (*R. Veeramani v The Secretary, Public Works Department*, 2012). Accordingly, the conduct of Committees at the appraisal stage needs to be strictly scrutinized for ensuring time-bound performance, transparency and impartiality in its every decision.

Judicial analysis of functioning of the Committees: grounds of judicial intervention

The scope of judicial interference in executive decisions has been demarcated on limited grounds (Jain and Jain, 2007, 1788). Intervention in these decisions cannot be imposed on their strict merit-based assessments but is instead focused on specific grounds, namely, illegality, procedural impropriety, irrationality and legitimate expectation (*Delhi Development Authority v M/s UEE Electricals Engg. Pvt. Ltd.*, 2004; *Jatinder Kumar v State of Haryana*, 2008; *State of NCT of Delhi v Sanjeev*, 2005).

Similarly, judicial intervention in the EC, being an executive decision by the Regulatory Authorities based on recommendations of Committees, can also only be based on the aforementioned grounds. Interestingly, the NGT has also declined to intervene in these decisions merely on grounds of the subjectivity of the opinions of Committee members and has limited interventions to questions of adequate consideration and appraisal of a project by Committees (*Anil Khedekar v Ministry of Environment and Forest*, 2014).

Based on readings of judgements on the EIA Notification, we have identified certain grounds that could form a basis for the NGT to intervene in Committee decisions under the EIA Notification.

Irregularities in the public consultation process

The EIA Notification is procedurally definite and unambiguous. The public consultation stage involves a set of mandatory procedural aspects to be followed by proponents and authorities before a project is sent for appraisal. Such aspects include: publicizing widely the public hearing for a project; furnishing accurate and adequate information to the public about the project; choosing a location for the hearing that is proximate to the project site; providing opportunity for the public to participate and raise concerns at the hearing and recording such concerns; and releasing minutes of the hearing. The rationale behind these steps is to ensure: effective participation of the public in this step in a free, fair and transparent manner; meaningful conduct of the public consultation and not as a mere procedural formality; and substantial involvement of the public in the sole step of the EIA process where they can participate. These aspects have made the public consultation stage a subject of rigorous judicial scrutiny (Mohan and Pabreja, 2016).

Although public consultation is mandatory in nature, not every procedural non-compliance signifies a significant procedural lapse in the EIA process (*Samata v Union of India*, 2014). Instead, only material (and non-curable) procedural deficiencies that cause material prejudice to the interests of local people or the environment are considered as grounds calling for a serious judicial response (*Adivasi Majdoor Kisan Ekta Sangthan v Ministry of Environment and Forests*, 2011).

Importantly, the EIA Notification requires the Committees to consider concerns raised at the public consultation stage before a final decision is taken (*Rudresh Naik v Goa State Coastal Zone Management Authority*, 2013). This requirement serves as a reality check for Committees and authorities about the practical implications of the project for local inhabitants and the environment/ecology of the region, and whether the project would serve its purposes as claimed by the proponents (*M P Patil v Union of India*, 2012). To meet this requirement, the Committees need to ensure during the appraisal that these aspects are meaningfully executed and that there is adequate compliance with the prescribed procedure [S.O. 1533, 2006, para 7(III)(iii)].

The failure of the EAC to consider the manner of execution of public hearing proceedings was considered by the NGT in *Adivasi Majdoor Kisan Ekta Sangthan v Ministry of Environment and Forests* (2011). The tribunal observed that failure of the EAC to consider material lapses in the conduct of the public hearing (being marred by violence, no reply furnished to concerns raised by public, etc.) during appraisal constituted a material violation of law, and, thus, the EC was set aside.

Similarly, in *Hanuman Laxman Aroskar v Union of India* (2019), the proponent failed to make fair and complete disclosure in the final EIA report about vital environmental concerns raised at the public consultation stage and effectively reduced them to a single issue before the EAC. It was held that the failure of the EAC to consider these aspects in detail amounted to dereliction of its prescribed duty and vitiated the process of appraisal.

The Committees are expected to exercise expert oversight over execution of various aspects of the public consultation stage in a meaningful manner. The concerns raised at the consultation stage and replies furnished thereon by the proponents provide the Committees with a deeper insight into the project. Accordingly, such lapses by them constitute material violation of the EIA Notification affecting the rights of the public in environmental decision-making.

Information in EIA reports

The information supplied in EIA documents forms the nucleus around which the whole EIA procedure revolves. The application for an EC for a project is accompanied with Form 1 (requiring proponents to submit information on various aspects related to the environment), which forms the basis for issuance of terms of reference by Committees for the proponents. These terms determine the scope of studies and form the basis for preparation by the proponents of the draft EIA report and ultimately the final EIA report. Such information – from Form 1 to the final EIA report – is crucial for adequate consideration of the project by Committees and effective implementation of the public consultation process (*Ossie Fernandes v Ministry of Environment and Forests*, 2011; *T. Murugandam v Ministry of Environment and Forests*, 2011). It is crucial that such information be adequate, accurate and authentic and provided in a timely and unambiguous manner (*T. Mohana Rao v The Director, Ministry of Environment and Forests*, 2012). Therefore, an appropriate assessment of such information by the Committees assumes huge significance.

In *Hanuman Laxman Aroskar v Union of India* (2019) – a case challenging the construction of a second international airport in Goa, the Supreme Court observed significant deficiencies in the information disclosed by proponents, including failure to note: (1) the existence of reserved forests, wetlands, water sources, biospheres and environmentally sensitive zones near the project site; (2) the impact of the proposed construction on birds and the flow of water in natural water channels; and (3) the number of trees to be felled for the project. The court emphasized that failure to disclose material information in Form 1 leads to the preparation of a deficient EIA report and, thus, has a cascading effect on salient objectives of the EIA Notification. Despite such critical remarks about the proponents and EIA consultants, the court merely suspended the EC and directed the EAC to revisit the proposal. Notably, the court did not also take action against the proponents, EIA consultants or responsible officials for deliberately concealing such critical information.

On the question of the relevance of the final EIA report for the appraisal of a proposal, the NGT observed that exhaustive and authentic information or data should be disclosed (*Jan Chetna v Ministry of Environment and Forests*, 2012; *Sunil Kumar Chugh v Secretary, Environment Department, Government of Maharashtra*, 2014). In a case involving incorrect information provided in an EIA report, the NGT pulled up the SEAC for negligence in not scrutinizing all relevant information before recommending the project (*Anil Khedekar v Ministry of Environment and Forest*, 2014). Similarly, it held an appraisal as vitiated due to a "consequent vacuum" created in the EIA report by the absence of material information related to bird species present on record in a region (*Save Mon Region Federation v Union of India*, 2012).

However, in another case, the NGT did not find absence of information like the date of report, date of visits and monitoring as a material inadequacy or deficiency. Instead of suspending the impugned EC, the tribunal directed the SEAC to seek clarifications from the proponent on these issues and submit a report to the MoEFCC as to the veracity of the EIA report (*Anil Khedekar v Ministry of Environment and Forest*, 2014). Similarly, the NGT justified the absence of the precise location of a project in the EIA report because concealment of this information could not be proved misleading or deliberate, and the EAC had adequately deliberated on the matter of location (*Ramesh Agrawal v Union of India,* 2013). Moreover, in an appeal challenging an EC for a river valley project, the NGT viewed inconsistencies among the terms of reference, EIA report and EAC minutes regarding land to be submerged and forest land to be diverted as the absence of material information for effective appraisal. Despite this, the tribunal refused to suspend the EC and instead constituted a committee to oversee the project (*Puran Chand v State of Himachal Pradesh*, 2012).

Based on the above, it can be inferred that deficiency or absence of information from the EIA report may not by itself become grounds for invalidity of an appraisal. Rather, for this to be the case, information absent from the report should be material to the EIA process or the absence should be misleading or deliberate. Though the EIA Notification expressly provides these grounds for rejection of a proposal [S.O. 1533, 2006, para 8(vi)], it is reasonable for the NGT to assess the absence of information in the context of the information's relevance for the EIA process. Pertinently, it raises the requirement for determining certain information in EIA documents as too crucial to be missing, incomplete or incorrect, where its absence or misstatement cannot be justified on grounds that it is not "material, deliberate or misleading." For instance, information about the period or date of site visits or monitoring is material for EIA, especially due to seasonal changes in ecology or seasonal presence of an endangered species in a region. Similarly, information on the precise location of a project in a draft EIA report is material for absolute clarity during public consultation. A mere executive deliberation on them might not be sufficient and be also contrary to the intent of the EIA Notification.

Another aspect worth noting is the critical view the NGT has taken of the lack of proper mechanisms in the EIA Notification for ensuring: (1) provision of authentic and adequate information in draft EIA reports (*Ramesh Agrawal v State Level Environment Impact Assessment Authority*, 2012; *Save Mon Region Federation v Union of India*, 2012); and (2) consistency of information between draft and final EIA reports. These shortcomings provide loopholes and can lead to provision of incomplete or incorrect information in EIA reports (*Ossie Fernandes v Ministry of Environment and Forests*, 2011). To rectify these problems, the NGT has consistently issued directions for authorities and proponents but has also reacted leniently towards subsequent non-compliances with these guidelines. In our opinion, these directions might continue to be irrelevant or ineffective unless incorporated into the EIA Notification or at least seriously considered by the judiciary itself while reviewing ECs, without which authorities would have no incentive to alter their course of action in future cases.

Importantly, the NGT's decisions in these cases might be considered an inadequate response. Others might consider them as an attempt to find a balance between the need to send a strong deterring message to proponents by suspending flawed ECs and the importance of such projects for economic and social development in the country. Nevertheless, on the basis of recurring non-compliances by government authorities and proponents, we believe that while these decisions might have saved these projects, their efficacy in creating deterrence among government authorities and proponents for future cases and ultimately safeguarding the environment is still uncertain.

Non-application of mind

Lack of "application of mind"[4] by the Committees on different questions related to a project becomes the grounds for judicial intervention in decisions taken under the EIA Notification. Such intervention is based on the assessment of discussions undertaken and reasons furnished by Committees before deciding on a proposal. The NGT has found it unreasonable to quash an impugned EC merely on a "hypothetical question, unfounded apprehension or non-scientific basis," unless reasons furnished by authorities shock the judicial conscience (*Husain Saleh Mahmad Usman Bhai Kara v Gujarat State Level Environment Impact Assessment Authority*, 2013).

Emphasizing on the role of the Committees in the process, the Delhi High Court commented that the *raison d'être* of Committees under the EIA Notification is to ensure an independent valuation of proposals. Thus, failure to provide reasons for recommendations would negate their very purpose, and even the subsequent application of mind by the MoEFCC on such proposals might not save unreasoned recommendations from being held as arbitrary decision-making and non-application of mind (*Utkarsh Mandal v Union of India*, 2009).

In an appeal challenging an EC for expansion of a port on the grounds that the appraisal process was faulty, the NGT observed that appraisal must involve an independent valuation of project documents by members of the EAC, which requires objective consideration of every proposal. Besides, their recommendations must be based on reasons that would assist other stake-holders as well as judicial authorities to adjudge if the EAC's decision was made after adequate application of mind. On the basis of these observations, the NGT suspended the EC after recording various instances of dereliction of official responsibilities by the EAC and the MoEFCC under the EIA Notification (*Gau Raxa Hitraxak Manch v Union of India*, 2012).

In *Conservation of Nature Trust v The District Collector, Kanyakumari District* (2013), the NGT directed the EAC to reconsider the proposal, as its neglect of studies presented for appraisal as well as material concerns raised during the public hearing had vitiated the process of appraisal. The tribunal noted that the EAC had failed to: appraise and apply its mind on concerns raised during the public hearing about the adverse implications of the project on numerous water bodies and trees in the region; seek clarifications on these issues from the proponent or conduct a site visit to assess real circumstances; and determine appropriate measures to mitigate impacts on water bodies or to minimize cutting of trees in the region.

A relevant source to adjudge application of mind by the Committees is the minutes of their meetings. The EIA Notification had this intent when it mandated the recording and release of these minutes on websites of Regulatory Authorities within five working days of the meeting (S.O. 1533, 2006, Appendix V, point 5). However, the nature of these minutes received a critical review by the NGT when it observed the recording of minutes in a generic, routine and stereotyped manner. It observed that for the sake of transparency in decision-making, these minutes should specify details of technical, sector and site-specific discussions that ensued in the meeting along with the "views, opinions, comments and suggestions" made by each member in a structured manner. Additionally, in case of failure to secure a unanimous decision, minutes must also specify reasons for the acceptance of the majority view. Thus, Committees were directed to ensure this in future appraisals (*Samata v Union of India*, 2014).

These discussions and reasons demonstrate the thought process of the EAC while deciding on a proposal. They serve as grounds on which the MoEFCC or judicial authorities can analyse whether the recommendations of Committees are based on "reasonable" consideration of all relevant factors related to a proposal. In one instance, the Supreme Court suspended an impugned EC on the basis of irrelevant reasons furnished by the EAC being: "peculiar circumstances of the case"; "larger public interest involved"; "no concealment of facts and circumstances of the case by proponents"; and "difficulties in land acquisition resulting in delay in preparation of EIA report." Since these reasons relied on extraneous considerations while neglecting crucial environmental considerations related to the project, the court directed the EAC to reconsider the proposal and revisit its recommendations (*Hanuman Laxman Aroskar v Union of India*, 2019).

These observations by the judiciary seem consistent with its limited intervention principle, that is, being cautious to intervene only in situations where it finds either absolute neglect of material concerns or extraneous considerations forming the grounds for the final decision. Through this approach, it has also emphasized the legislative intent for establishing the Committees – consisting of members with "experience or expertise" in the environment – to perform their functions without unwarranted interference from other institutions.

Inadequate consideration

Committees have the responsibility to undertake detailed consideration of project-related documents and reports (*Gau Raxa Hitraxak Manch v Union of India*, 2012). The scope of consideration during appraisal cannot be restricted to only information provided in the EIA report. Rather, a Committee's scope is wide enough to consider almost all aspects of the project, including technical and legal aspects, adequacy and authenticity of material information in EIA reports, concerns raised at public consultation, responses submitted by the proponent to questions posed by the EAC (*Goa Foundation v Goa State Environment Impact Assessment Authority*, 2015), adequacy of environmental management plans and conditions or safeguards imposed on the proponent for mitigation of environmental impacts. Given their significance, adequacy of consideration of these issues by Committees forms a relevant criterion for the judiciary to adjudge the validity of the appraisal step.

Highlighting lacunae in consideration of material issues related to a proposal, the NGT suspended the EC granted to a proposal concerning offshore oil and natural gas extraction on three grounds. First, the SEAC neglected the MoEFCC Office Memorandum that required such proposals to be appraised after obtaining recommendations from the State Coastal Zone Management Authority concerned. Second, the appraisal was conducted by the SEAC without considering relevant factors, namely, selection of site, changes (if any) in the original plan, status of construction of the project, impact on the environment and need for a rapid environment report based on terms of reference in the EIA studies. This omission occurred despite the MoEFCC's previous communications and the Maharashtra Pollution Control Board's stop work order against the proponent for commencing the impugned project without requisite clearances. Third, the SEAC acted unreasonably when it removed a condition, which barred the proponent from initiating land reclamation in the project site, from EC recommendations without considering the environmental implications (*Naim Sharif Hasware v Das Offshore Engineering Pvt. Ltd.*, 2014).

Similarly, another pertinent issue deliberated by the NGT was the validity of an EC whose appraisal step had been marred by the neglect of information on the project site as well as a highly irregular public hearing that included a threat of violence against members of the public. The NGT observed that the EAC failed to: consider material information on the nature of the land, the presence of forestland and wildlife, and mineral resources within and around the project site; assess the situation on the ground; and conduct a site inspection along with field experts. Furthermore, it ignored concerns raised by the Banaras Hindu University regarding potential contamination of water reservoirs by effluents released from the project into the Ganga River. These omissions were exacerbated by inconsistencies between information provided by proponents in the EIA report and responses received by the appellants on a right to information query. However, despite these inconsistencies and neglect, the NGT merely suspended the EC without holding either the proponent or the EAC responsible for material lapses in the information supplied and the appraisal process, respectively (*Debadityo Sinha v Union of India*, 2014).

In another case, the NGT discussed the conduct of the SEAC in recommending an EC for a proposal when the proponent had not only commenced the project without a clearance but also provided an inadequate environmental management plan for the SEAC's consideration. While the SEAC justified its positive recommendations with future rectification of these shortcomings by the proponent, the NGT found that mandatory conditions for the EC had been blatantly violated by the proponent. Still, the NGT did not set aside the EC but provided additional safeguards and levied compensation for restitution of environmental damage caused by the proponent (*Sunil Kumar Chugh v Secretary, Environment Department, Government of Maharashtra*, 2014).

It is difficult to justify this order, as it effectively provides judicial sanction for improper appraisal by the SEAC as well as commencement of a project by the proponent before a clearance is granted. In this context, mere compensation for restitution for damages to the environment is not an adequate judicial response. This "violate first–pay later" approach of the tribunal, if used in cases of blatant non–compliance, would establish an ill–conceived precedent inconsistent with the precautionary principle as well as the objectives of the EIA Notification. Accordingly, the NGT should have taken a strong stance against this substantial lapse of the SEAC in performing its official duties and suspended the EC along with imposing a penalty against erring persons.

In another case, the NGT deliberated on the relation between adequate consideration and appropriate time to be allotted for each proposal by the EAC in its meetings. This issue arose when the appellants argued that the EAC had rushed through its agenda because it had to deal with a large number of proposals in a single day. Rejecting this argument, the NGT stated that though the EAC has to carefully scrutinize every proposal, it still has the discretion to determine the appropriate time for each proposal. As such, the NGT cannot determine a time frame that the EAC should allot for each proposal (*M.D. Yogesh v Union of India*, 2016). However, a similar claim related to an appraisal done in a "hasty, cursory and arbitrary manner" became grounds to suspend an EC granted to an international airport in Kerala when the NGT found that the proposal had been considered along with a long list of other proposals in the same meeting (*Sreeranganthan K.P. v Union of India*, 2013). Though a time frame cannot be specified for consideration of each project by the EAC, hasty and evasive conduct at the cost of reasonable appreciation of

the environmental implications of a project would negate the EAC's fundamental purpose under the EIA Notification (*Samata v Union of India*, 2014).

Pertinently, deliberations done by the EAC give a substantive meaning to
the EIA process, and reasons furnished in this process serve as a check for
authorities to ensure that the decision was taken in a non-arbitrary manner
(Menon and Kohli, 2015). The relevance of inadequate consideration of material aspects of a project can be understood from the impacts that Committees'
lapses can have on the environment and communities of a region. At the
least, they might result in the approval of a project, the reports of which were
based on logically inconsistent or misleading information. Alternatively, they
might be based on misleading information related to the nature of land to be
acquired for the project, or about technology to be utilized for the project.
This inadequate information and ignorance of this fact by the Committees
or Regulatory Authorities might also adversely alter the safeguards and conditions appropriate for a project. Thus, it is not hard to imagine the adverse
consequences of such approvals, primarily on the basis of the precautionary
principle which lies at the heart of the whole EIA procedure.

Conclusion

A major reason for global acceptance of sustainable development as a conceptual solution in the fierce "development versus environment" or, to be more
precise, "privilege versus livelihood" debate has been the term's capacity to
intertwine two fundamental concepts of environmental jurisprudence, the
precautionary principle and the polluter pays principle. While the almost
irreversible nature of actions affecting the environment would demand more
emphasis on the former principle than the latter, actions on the ground reflect
a contrary picture – with projects having adverse environmental impacts
being allowed without adequate caution, essentially in the name of development. The inclination towards *ex post facto* clearances makes it worse. It
is only after the cost of unregulated development and the limited nature of
natural resources become apparent that governments realize the need to alter
their course of action.

In the Indian context, the EIA Notification was intended to be the mechanism for such a change. It sought to bring together the legislature, the
executive branch and the public to deliberate over these issues and achieve
environmental justice. The multidisciplinary constitution of Committees
and the credibility of their recommendations before Regulatory Authorities
manifest their significant role in environmental decision-making. Given
that the entire EIA procedure revolves around the Committees, their role
in securing environmental justice for different stakeholders assumes huge
significance.

Accordingly, the judicial concerns with respect to Committee constitution and manner of decision-making have been given a special emphasis in
this chapter. On the basis of our reading, we observed that the judiciary has

rightly adhered to the principle of limited judicial intervention in executive decisions. However, its intervention in decisions that are inconsistent with fundamental aspects of environmental law has indeed been required. In such cases, the judiciary has consistently reminded proponents and authorities of their duties and essentially ordered reappraisal of projects; suspended ECs granted to projects; or issued guidelines to streamline future conduct.

However, in our opinion, while these responses might be justified in some instances, a general proclivity towards such responses could set a wrong precedent and may not bode well for the quality of future environmental decisions by such authorities. This observation holds true especially for instances where the NGT decides to save an EC even when it finds information provided for the EC to be incomplete, inconsistent or lacking certain fundamental aspects, or where authorities have been blatantly derelict in their statutory duties. In this regard, suspension of impugned ECs alone does not seem an appropriate response, as it may not really improve the quality of the process in future. Thereby, we argue that the judiciary needs to evolve a pragmatic balance between the need to save a project and the importance of communicating to all stakeholders its seriousness towards the sanctity and ultimate objective of the EIA process – which at the same time also ensures that proponents, Committees or Regulatory Authorities assume accountability for their actions and perform their duties with utmost diligence.

In our opinion, besides issuing guidelines, directing reappraisal of the project or suspending the EC, an appropriate response to serve this purpose would be holding the delinquent proponents, Committees and Regulatory Authorities directly responsible for their actions in cases where there are serious lapses on their part. This response would provide the required balance, as it would allow the judiciary to acknowledge non-material or curable defects in the EIA process where another opportunity can be given (through directing reappraisal, suspension of the EC, or issuing guidelines) if they are not severe, deliberate or intended to deceive. However, in instances involving material or incurable defects, a lenient response might make proponents or authorities non-diligent towards their obligations under the EIA process or allow them to take the EIA process for granted, since they would know in advance that the judiciary would not resort to harsh actions against them (or hold them accountable) even in cases of material defects. The judiciary would have to determine what constitutes material or curable defects based on specific facts and circumstances.

Moreover, we also observed that Committees are required to consider a large number of projects within the appraisal time prescribed in the EIA Notification. Given their limited number of meetings, it is evident that Committees are unable to give adequate time to each proposal, thus naturally affecting the quality of appraisal in the EIA process. In this respect, we believe that Regulatory Authorities need to ensure that there are adequate numbers of sector-specific Committees to reduce the burden on a few members. At the same time, Committees should ensure that they meet more regularly according to the number of pending proposals – which will improve the quality of appraisal and comply with statutory requirements.

Additionally, in relation to the quality of EIA documents, it is of utmost importance that proper mechanisms be in place for ensuring the adequacy, accuracy and veracity of the information provided in EIA documents. There have been numerous instances of information in EIA documents being found misleading or incomplete, where proponents have been excused from this essentially on the grounds of the information not being "material, deliberate or misleading" and/or adequate executive deliberation over the appropriate information. However, this approach becomes problematic when the information in question is absolutely crucial for the proposed project, particularly for the public consultation stage. In light of this, we opine that certain information in EIA documents (depending on the nature of the project) should be earmarked as absolutely crucial for consideration of the application for an EC. Importantly, concealment, absence or misstatement of such information cannot be justified on grounds that it is not "material, deliberate or misleading" or that appropriate information has been considered by Committees at the appraisal stage.

That said, after almost fifteen years of the enactment of the EIA Notification, we believe that a course correction is in order for bodies responsible for its execution as well as the judiciary overseeing the whole process. It is also the seriousness with which the government responds to these judicial opinions that would signify whether it is even remotely interested in implementing the EIA Notification in letter and spirit.

Notes

1 Touted as one of the world's worst industrial disasters, in 1984, Bhopal city in India witnessed the death of thousands of people after toxic gas was leaked from a factory owned by Union Carbide Corporation situated in the residential area of the city.
2 At the time of writing this chapter, the Union Government had invited public comments to the draft version of the EIA Notification 2020. However, the final version had not been notified yet.
3 The National Green Tribunal is a specialized tribunal created by the Parliament of India to specifically deal with environmental matters.
4 Generally used in legal terminology, this phrase refers to whether an administrative authority has given adequate consideration to all relevant aspects related to a proposal before making the final decision.

References

Adivasi Majdoor Kisan Ekta Sangthan v Ministry of Environment and Forests, Appeal No 3 of 2011 (National Green Tribunal).

Anil Khedekar v Ministry of Environment and Forest, Appeal No 38 of 2014 (WZ), order dated 14 July 2017 (National Green Tribunal).

Banerjee, S. (2015) "Experts without expertise" (blog), *Down to Earth*, 31 August. Available at: https://www.downtoearth.org.in/coverage/experts-without-expertise--45800 (Accessed: 22 May 2021).

Cole, L. and Foster, S. (2001) *From the Ground Up: Environmental Racism and the Rise of the Environmental Justice Movement*. New York, NY: NYU Press.

Conservation of Nature Trust v The District Collector, Kanyakumari District, Application No 104 of 2013 (SZ), order dated 14 September 2016 (National Green Tribunal).

Debadityo Sinha v Union of India, Appeal No 79 of 2014, order dated 21 December 2016 (National Green Tribunal).

Delhi Development Authority v M/s UEE Electricals Engg. Pvt. Ltd. (2004) 11 SCC 213.

Foster, S. (2002) "Environmental justice in an era of devolved collaboration," *Harvard Environmental Law Review*, 26, 459–498.

Garner, B. (2019) "Appraisal" in *Black's Law Dictionary*. 11th edn. Toronto: Thomson West.

Gau Raxa Hitraxak Manch v Union of India, Appeal No 47 of 2012, order dated 11 August 2013 (National Green Tribunal).

Ghosh, S. (2013) "Demystifying the environmental clearance process in India," *NUJS Law Review*, 6(3), 433–480.

Goa Foundation v Goa State Environment Impact Assessment Authority, Application No 135 of 2015 (WZ), order dated 29 November 2016 (National Green Tribunal).

Hanuman Laxman Aroskar v Union of India (2019) 15 SCC 401.

Husain Saleh Mahmad Usman Bhai Kara v Gujarat State Level Environment Impact Assessment Authority (2013) SCC OnLine NGT 81.

Jain, M. and Jain, S. (2007) *Principles of Administrative Law*, vol. 2. 6th edn. New Delhi: LexisNexis India.

Jan Chetna v Ministry of Environment and Forests (2012) SCC OnLine NGT 81.

Jatinder Kumar v State of Haryana (2008) 2 SCC 161.

Jenkins, B. (2018) *Water Management in New Zealand's Canterbury Region: A Sustainability Framework*. Dordrecht, The Netherlands: Springer.

Kalpavriksh v Union of India, Application No 116 (THC) of 2013, order dated 17 July 2014 (National Green Tribunal).

Lower Painganga Dharan Virodhi Sangharsha Samiti v State of Maharashtra, Application No 13 (THC) of 2013 (WZ), order dated 10 March 2014 (National Green Tribunal).

M P Patil v Union of India, Appeal No 12 of 2012 (National Green Tribunal).

M.D. Yogesh v Union of India, Appeal No 121 of 2016 (SZ), order dated 13 February 2017 (National Green Tribunal).

Menon, M. and Kohli, K. (2009) "From impact assessment to clearance manufacture," *Economic and Political Weekly*, 44(28), 20–23.

Menon, M. and Kohli, K. (2015) "Environmental regulation in India: Moving 'forward' in the old direction," *Economic and Political Weekly*, 50(50), 20–23.

Mohan, M. and Pabreja, H. (2016) "Public hearings in environmental clearance process: Review of judicial intervention," *Economic and Political Weekly*, 51(50), 68–75.

Naim Sharif Hasware v Das Offshore Engineering Pvt. Ltd., Application No 15(THC) of 2014, order dated 24 December 2014 (National Green Tribunal).

Ossie Fernandes v Ministry of Environment and Forests, Appeal No 12 of 2011 (National Green Tribunal).

Puran Chand v State of Himachal Pradesh, Appeal No 48 (THC) of 2012, order dated 2 February 2016 (National Green Tribunal).

R. Veeramani v The Secretary, Public Works Department, Appeal No 31 of 2012, order dated 20 February 2013 (National Green Tribunal).

Ramesh Agrawal v Union of India, Appeal No 08 of 2013 (CZ), order dated 22 August 2014 (National Green Tribunal).

Ramesh Agrawal v State Level Environment Impact Assessment Authority (2012) SCC OnLine NGT 32.

Rudresh Naik v Goa State Coastal Zone Management Authority (2013) All (1) NGT Reporter (2) (Delhi) 47.

S.O. 1533, Ministry of Environment and Forest Notification (updated copy dated August 2015). Available at http://environmentclearance.nic.in/writereaddata/public_display/circulars/56_UniEIANoti2006.pdf (Accessed: 22 May 2021).

S.O. 1533, Ministry of Environment and Forest Notification (14 September 2006).

S.O. 1737 (E), Ministry of Environment and Forest Notification (11 October 2007). Available at: http://environmentclearance.nic.in/writereaddata/EIA_Notifications/3_SO1737E_11102007.pdf (Accessed: 22 May 2021).

Samata v Union of India (2014) 1 All India NGT Reporter (SZ) (National Green Tribunal).

Save Mon Region Federation v Union of India, Appeal No 39 of 2012 (National Green Tribunal).

Schlosberg, D. (1999) *Environmental Justice and the New Pluralism: The Challenge of Difference for Environmentalism*. Oxford: Oxford University Press.

Sreeranganthan K.P. v Union of India, Appeal Nos 172–174 of 2013.

State of NCT of Delhi v Sanjeev (2005) 5 SCC 181.

Sunil Kumar Chugh v Secretary, Environment Department, Government of Maharashtra, Appeal No 6 of 2014, order dated 3 September 2015 (National Green Tribunal).

T. Mohana Rao v The Director, Ministry of Environment and Forests (2012) SCC OnLine NGT 40.

T. Murugandam v Ministry of Environment and Forests, Appeal No 17 of 2011, order dated 23 May 2012 (National Green Tribunal).

Utkarsh Mandal v Union of India (2009) (10) AD (Delhi) 365.

Yang, T. (2002) "Melding civil rights and environmentalism: Finding environmental justice's place in environmental regulation," *Harvard Environmental Law Review*, 26, 4–8.

Part III

Cases studies of justice and injustice

7 Small hydro and environmental justice

Lessons from the Kullu District of Himachal Pradesh

Alan P. Diduck, Richard Johnson, Esther Edwards,
A. John Sinclair, James Gardner and Kirit Patel

Introduction

Himachal Pradesh has significant hydropower potential, and the public sector (state) is rapidly developing this resource in conjunction with private developers (Kumar and Katoch, 2014), via state-based agencies (e.g. Himachal Pradesh Energy Development Agency [Himurja]). Because of the negative impacts associated with large-scale projects, Himachal Pradesh has promoted small hydro, which is thought to offer more potential for social and economic benefits for local residents and fewer adverse impacts (Sharma, Tiwari and Sood, 2013). Although the Government of India defines small hydro as projects with less than 25 MW of capacity, Himachal Pradesh defines small hydro as those less than 5 MW (Himurja, 2015b; Mishra, Khare and Agrawal, 2015).

India's national impact assessment legislation is enacted by a regulation passed under the Environment (Protection) Act 1986. Furthermore, the Environment Impact Assessment (EIA) Notification 2006 (S.O. 1533, 14 September 2006) applies to hydropower projects but includes important exemptions. Assessment is compulsory for projects greater than 50 MW, while projects between 25 and 50 MW are screened to determine if an assessment is necessary, and a formal assessment is not required for projects smaller than 25 MW (Erlewein, 2013; EIA Notification 2006 Schedule 1).

The regulatory gap created by the exemption for small hydro has been filled in part by state approval processes. In Himachal Pradesh, project proponents are required to prepare detailed reports describing geological, hydrological, engineering and financial aspects of their projects, along with the economic, environmental and social impacts. They are also required to apply for a No Objection Certificate (NOC) from village councils (Gram Panchayats) if local communities will be affected. Until 2014, NOCs also were required from relevant government ministries, such as Public Works, Irrigation and Public Health and Fisheries and Wildlife (Himurja, 2015c).

DOI: 10.4324/9781003141228-7

Given the strategic importance of small hydro development in Himachal Pradesh and the adverse impacts associated with rapid growth of that sector (e.g. disruption to local irrigation systems, loss of traditional livelihood opportunities and loss of cultural assets) (Baker, 2014c; Diduck and Sinclair, 2016; Rai and Srivastava, 2014), the objectives of this research were to (1) examine the impacts of small hydro development in Himachal Pradesh and (2) consider the environmental justice implications for local communities.

Small hydro in Himachal Pradesh

Himachal Pradesh, located in the Western Himalaya, holds considerable hydropower potential, estimated to be 27,463 MW, of which 24,000 can be harnessed. In 2017, the state had an installed capacity of 10,519 MW (Government of Himachal Pradesh, 2020). The state views hydropower as a key driver of economic growth, and in recent years, it has accelerated the pace of hydropower development (Kumar and Katoch, 2014; Government of Himachal Pradesh, 2020). Because of the negative impacts associated with large-scale projects and to take advantage of smaller sources, Himachal Pradesh has actively promoted small hydro by adopting policy measures, such as wheeling and banking of power, streamlined clearance processes and waiver of royalties for designated periods (Baker, 2014a; Diduck and Sinclair, 2016; Chapter 2, this volume). Statewide, by 2020 there were 742 small hydro projects that had been allotted, with an aggregate capacity of nearly 1,779 MW. Therein, 281 implementation agreements had been signed, of which 161 projects were at the clearance stage, 88 had been commissioned and 32 were under construction (Government of Himachal Pradesh, 2020).

In Kullu District alone, 137 small hydro projects had been allotted and 70 implementation agreements signed by 2015 (45 projects were in clearance, 12 had been commissioned and 13 were under construction) (Himurja, 2015a). The Kullu District is a high mountain region, with altitudes ranging from c. 1,000 to 6,600 m above sea level (Johnson, Edwards, Gardner and Diduck, 2018). Pertinent to hydropower generation, it has substantial stream/river flows derived primarily from glacier melt, snowmelt and rainfall in the June–September monsoon season (Sah and Mazari, 2007). The district had a resident population of 440,000 in 2011, of which 90 per cent lived in rural areas (Johnson, Edwards, Gardner and Mohan, 2015), and a substantial transient and migratory population. Economic activity in the district includes agriculture, horticulture and aquaculture; tourism and pilgrimage; hydroelectric development; cottage and small industries; and local commerce (Directorate of Census Operations Himachal Pradesh, 2011).

As noted, it is thought that small hydro offers more potential for social and economic benefits for local residents and generally has fewer adverse impacts (Sharma, Tiwari and Sood, 2013; Mishra, Khare and Agrawal, 2015). However, the research literature reveals that predicted or promised local benefits such as employment, improved road networks, more reliable power and reduced combustion of wood often are not realized, and impacts are far from benign (Kumar and Katoch, 2015; McCandless, 2007; Sinclair, 2003; Sinclair, Diduck and McCandless, 2015). Studies from Himachal Pradesh, including the Kullu District, have demonstrated the range and severity of adverse impacts experienced during construction and subsequent scheme operation. These impacts include disruption of local irrigation systems, loss of traditional livelihood opportunities, loss of cultural assets, removal of trees and horticultural land, diverted channel flows and altered sediment transport regimes, slope instability, soil erosion and increased flood risks – many of which are manifest on site and downstream (Baker, 2014a; Diduck and Sinclair, 2016; Rai and Srivastava, 2014; Chapter 2, this volume).

Despite increased understanding of adverse impacts, small hydro projects are not subject to the EIA Notification 2006 and therefore do not fall under the purview of the National Green Tribunal (NGT) (the NGT Act, Schedule 1). Furthermore, small hydro is not subject to formal approval processes by relevant state government ministries and agencies, such as Public Works, Irrigation and Public Health, Fisheries and Wildlife and the Disaster Management Authority (Himurja, 2015c). Project proponents are required to prepare a Detailed Project Report (DPR) describing geological (but not dedicated geomorphological appraisal), hydrological, engineering and financial aspects of their projects, along with economic, environmental and social impacts. However, the DPRs are site specific and do not fully assess connected and coupled impacts beyond and between sites. Proponents are also required to provide 1 per cent of the project cost as "local area development funds," which are meant to support local infrastructure and other development projects, as well as apply for NOCs from affected Gram Panchayats. Beyond this, there are no formal requirements to involve the public in preparation of the DPR or to otherwise engage in broad community consultations.

Theoretical orientation

This chapter adopts the framework of environmental justice found in Chapter 1. The framework is founded on basic attributes of good governance, such as broad-based citizen participation, respect for the rule of law, respect for cultural values, effectiveness and efficiency, accountability, transparency and responsiveness (United Nations Economic and Social Commission for Asia and the Pacific, 2008). The framework is also informed by the goals of sustainable development, such as inter- and intragenerational equity (World Commission on Environment and Development, 1987) and United Nations Sustainable Development Goal 16. This goal is to "promote peaceful and inclusive societies for sustainable development, provide access to justice for all and build effective, accountable and inclusive institutions at all levels" (United Nations Department of Economic and Social Affairs, 2020).

Included in the framework are four basic components. Recognitional justice refers to recognition of the diversity of participants, experiences and interests (desires, wants, needs, goals and aspirations) in communities that are affected by environmental governance decisions (Schlosberg, 2004; Williams and Mawdsley, 2006). This component provides a basis for and enables the other three components. Procedural justice requires opportunities for meaningful participation in environmental governance and the political and legal processes to create and manage them (Gill, 2017; Pring and Pring, 2009). Meaningful participation in governance and political processes requires early and ongoing opportunities for involvement, access to information and, if necessary, access to resources to gain the capacity for effective participation, among other key features (Stewart and Sinclair, 2007; Diduck and Sinclair, 2021). Meaningful participation in legal processes requires access to information, knowledge and legal and technical support, along with legal standing; access to proceedings that are fair, efficient and affordable; and access to enforcement mechanisms (Pring and Pring, 2009).

Distributive justice, which is at the heart of the framework, seeks equity in the distribution of the risks and benefits that result from environmental governance decisions (Schlosberg, 2004; Williams and Mawdsley, 2006). Finally, restorative justice is concerned with the extent to which the adverse impacts of environmental governance decisions are avoided, mitigated and remedied (Motupalli, 2018; Raphael, 2019).

Research strategy and methods

Our research strategy involved a preliminary scoping phase followed by an intensive examination of five small hydro projects in Himachal Pradesh. Methods included document review, field observations and semi-structured interviews. Data sources for the document review were public records, such as legislation, policies, government reports and project-specific approval documents. Field observations were recorded with maps, notes and photographs. The interview participants included farmers, shop owners, village leaders, community activists, members of conservation and other community organizations, other local residents, state government officials and project proponents or owners and their employees.

The scoping phase included 20 interviews that canvassed small hydro growth in Himachal Pradesh, as well as 15 projects[1] that were considered but not selected for the intensive phase of the research. The intensive phase, which involved 32 interviews, centred on the Chorr, Haripur, Kathi, Kukri and Pakhnoj projects in the Beas River watershed in the Kullu District (Figure 7.1). These projects were chosen because of similar design (run-of-the-river) and initiation in the private sector, the availability of background documentation and a significant local interest in them.

The interviews were conducted during four field visits (July 2012, October 2013, April 2014 and April 2015). We used non-probability, purposive and snowball sampling (Creswell, 2014) to recruit interview participants. Some interviews were conducted in English, but most required an interpreter. In

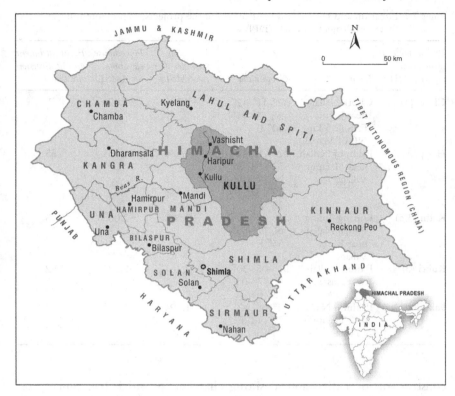

Figure 7.1 Map of the Kullu District in the Beas River watershed. The Chorr project is near Vashist village, Haripur and Pakhnoj are near Haripur village, and Kathi and Kukri are approximately 30 kilometres north-northwest of Kullu town.

all instances, the interview data were recorded in field notebooks or digitally recorded in accordance with ethics approvals received beforehand.

The interview questions focused on small hydro growth, project approvals, local participation in project development and understandings of project impacts. Data analysis was inductive, began in the field and continued in the laboratory using QSR NVivo 11 qualitative data analysis software. The analysis took a grounded approach (Corbin and Strauss, 2014) and involved sorting and coding data segments and identifying themes (Creswell, 2014). Throughout, efforts were made to verify the data with field notes and/or documentary data.

Results

The five run-of-the-river projects

The projects were designed as run-of-the-river developments with hydraulic structures to divert water to a small powerhouse. Such projects involve

Table 7.1 Location and selected salient features of the projects, as presented in the Detailed Project Reports (DPRs)

Project name (generation capacity, MW)	Location	Coordinates	Diversion weir elevation (m AMSL)	Powerhouse elevation (m AMSL)	Basin catchment area at diversion site (km²)
Chorr (1)	Chorr Nallah, near Jogini Falls (2.5 km NE of Manali)	32°-11'-42" N; 77°-11'-40" E	~2,280	~2,078	7
Haripur (3)	Pakhnoj Nallah, near Haripur Village	32°-08'-44", 32°-08'-32" N; 77°-10'-29", 77°-09'-33" E	~1,686	~1,531	35
Kathi (3.5)	Umang/Joling/ Phojal Nallahs, near Kasta Village	32°-08'-10" N; 77°-03'-55" E	~2,382	~2,003	18
Kukri (5)	Balsoti Nallah, near Kasta Village	32°-05'-20" N; 77°-05'-33" E	~2,461	~2,220	52
Pakhnoj (2.5)	Pakhnoj Nallah, near Haripur Village	32°-08' N; 77°-12' E	~1,994	~1,710	52

construction of a diversion, desilting chamber, penstock (i.e. sluice/gate/ pipeline/intake structure controlling water supply to the turbine), power-house and tailrace (i.e. a channel carrying used water away from the turbine). Further technical details of run–of–the–river projects can be found in Chapter 2. Table 7.1 summarizes the location and selected salient features of the projects as presented in the DPRs.

Grounded thematic analysis

The thematic analysis of the interviews yielded five primary themes: adverse impacts (perceived, experienced and expected); benefits (perceived, experienced and expected); governance and approval processes; public participation (or lack thereof) in governance and approval processes; and community opposition. Each primary theme included secondary and, in some cases, tertiary themes. Table 7.2 identifies the themes and, as a meas-ure of their predominance, includes the number of sources (interviews) that expressed each theme and the number of references (data segments) coded to each theme. The themes are explained below, and selected inter-view quotations are provided to add depth and richness of detail. Code names are used in the quotations to protect the anonymity of the research participants.

Table 7.2 Themes derived from the interviews, along with the number of sources
(interviews) and *references* (data segments) coded to each theme

Primary themes	Secondary themes	Tertiary themes
Adverse impacts (42, *192*)	Water (27, *90*), social and economic (16, *48*), terrestrial systems (15, *19*), minimal adverse impacts (9, *13*)	*Water*: agriculture (19, *33*), human uses (16, *20*), fish stocks (8, *17*) *Social and economic*: sacred spaces (10, *34*), economy (5, *7*), personal safety (4, *4*) *Terrestrial systems*: landslides and soil erosion (9, *10*), trees and plants (7, *9*)
Benefits (35, *91*)	Local area development funds (19, *44*), minimal benefits (17, *22*), job creation and economic spinoffs (17, *21*)	*Local area development fund*: infrastructure (11, *20*), public services (6, *9*)
Governance and approvals (26, *47*)	State government (17, *24*), panchayat (7, *7*), project proponents (6, *9*), remediation (5, *7*)	
Public participation (or lack thereof) (30, *67*)	Minimal involvement (26, *36*), Panchayat NOC concerns (17, *31*)	*Minimal involvement*: lack of opportunities (15, *19*), lack of adequate notice (12, *12*), ways to improve involvement (4, *5*) *Panchayat NOC concerns*: not granted (10, *13*), deceitfully attained (9, *18*)
Community opposition (19, *54*)	Public protests (13, *23*), court challenges (11, *19*), community-based organizations (10, *12*)	

Adverse impacts

Adverse impacts (perceived, experienced and expected) formed the predominant primary theme (42 sources, *192 references*), which included four
secondary themes. Water was the most common secondary theme (27, *90*),
consisting of three tertiary subthemes spanning concerns about agriculture
(irrigation, rice production and livestock) (19, *33*), human uses (availability
of potable supplies) (16, *20*) and fish stocks (both wild and farmed) (8, *17*), as
expressed by the following comments:

> The main issue is the availability of water. We have horticulture such as
> apple orchards. We do not get enough water for irrigation.
>
> (Ravi, farmer and gharat owner, Pakhnoj project)

> I am concerned about the pollution of water. I am afraid that people will
> have to drink polluted water.
>
> (Rubina, local resident, Kukri project)

> It is going to affect the fish farm owners in this area. There are five fish
> farms and mostly those are private and all of them are using the water
> from the stream.
>
> (Manohar, shop owner, Haripur project)

The next most common secondary theme identified social and economic impacts (16, *48*), highlighting three tertiary subthemes: disruption of sacred spaces (10, *34*), economic harm from reduced tourism and closure of gharats (traditional watermills) (5, 7) and threats to personal safety from construction crews (4, *4*). Disruption of sacred spaces was mainly a concern in the Chorr case, as the proposed project would have impacted two temples to the goddess Jogini and a waterfall named in her honour:

> Stones near the base of the falls were considered to be the footprints of Jogini, and many were broken during construction. At one time no one was allowed to go there, even pujaris [temple priests] could only enter on holy occasions, although some exceptions were made for pastoralists for grazing.
>
> (Ajay, teacher, Chorr project)

The secondary theme of terrestrial systems (15, *19*) included concerns about landslides and soil erosion (9, *10*) and loss of trees and plants (7, *9*). Notably, and counter to the foregoing concerns, nine research participants, including project proponents, government officials and local residents alike, thought there were minimal adverse impacts on terrestrial systems (9, *13*).

> There are no negative effects of the project. Only a few trees will be cut if they are bringing the pipeline. But the rest there is no problem.
>
> (Anupam, local resident, Kathi project)

> I think there is no impact of small hydro projects. There are issues related to big projects, such as construction of infrastructure, land issues, submergence of areas, cutting trees, etc. But in the case of small hydro projects, these are really too small to have any impacts.
>
> (Vishal, state government official)

Benefits

Benefits (perceived, experienced and expected) made up the second most common primary theme (35 sources, *91 references*) and comprised three secondary themes. Local area development funds, the most common of the secondary themes (19, *44*), encompassed suggestions and ideas about funding provided by the project proponent – equalling 1 per cent of project costs – to the Local Area Development Authority in support of development activities. Two notable tertiary themes were infrastructure (11, *20*), such as streetlights, paved roads and paths, bridge repairs, sanitation facilities and temple upkeep and public services (6, *9*) such as more reliable electricity and improved education and medical services.

Job creation and economic spinoffs (17, 21) formed the third most common secondary theme. The latter included tax revenue, increased business for hotels and shops from migrant workers and potential donations for festivals. Job creation, and in particular employment in project construction, was commented upon by local residents and project opponents as well as by project employees, as for example:

> There are many unemployed youths in these three panchayats. If through the project they are getting some kind of employment, then that will be helpful.
>
> (Prem, opposition leader, Haripur project)

> People are making money because they are getting jobs, so they are getting Rs. 8,000–Rs. 10,000 per month. So, people would also like to have jobs in the other villages.
>
> (Anupam, local resident, Kathi project)

> There are 10 or 15 people who have got employment right now, and when this project was being built at that time there were about 60 or 70.
>
> (Ankush, employee, Suman Sarwari project)

In contrast to these positive views, an important secondary theme (17, *22*) reflected positions of people who saw minimal or no benefits associated with small hydro:

> We do not get any benefits out of these projects. It may be beneficial to the country but definitely not for the local people.
>
> (Ravi, farmer and gharat owner, Pakhnoj project)

> No benefits. They are not giving the electricity to any of the villages. Maybe they will give the electricity to the government, so what kind of benefit are we going to get?
>
> (Purva, village leader, Haripur project)

> Yes, [there are jobs] during construction. After that there is no employment, no continuous employment. When you complete a hydro project then you need only very few people and they are all experts.
>
> (Vijay, conservation organization)

Governance and approvals

The third and fourth primary themes, governance and approvals (26 sources, *47 references*) and public participation (30, *67*), are closely linked and had similar prevalence in the NVivo database. The third theme is summarized here and the fourth in the next section. Governance and approvals included three secondary themes, encompassing descriptive information on the roles played by state government agencies and departments (17, *24*), panchayats (i.e. village councils) (7, 7) and project proponents (6, *9*) in preparing, reviewing and approving project proposals, as the following comments show:

> Clearances and NOCs are required from agencies with jurisdiction over affected areas, such as the Forest Department, the panchayat, the Revenue Department, Sub-Divisional Magistrate or the District Commissioner, Irrigation and Public Health, the Public Works Department, Fisheries Department, and Pollution Control Board.
>
> (Satyananda, state government official)[2]

> In the panchayat there are different types of people. Some have got a personal interest, but the overall interest is for the betterment of the area. If the panchayat wanted, they can object right from the beginning that this project should not be established because it will cause more harm, it is more harmful than beneficial.
>
> (Virbhadra, opposition leader, Haripur project)

> Our project was self-identified, but we still used the standard application process. All documents were looked at first by Himurja before implementation. The documents had to be collected and organized by us, all NOCs had to be obtained by us, and coordinated by us. We had to speak with all separate government ministries after receiving the requirements from Himurja.
>
> (Dev, project owner)

Finally, governance and approvals included a secondary theme pertaining to remediation (5, 7), following the lack of restoration and clean-up in the Chorr case after the project was halted because of public opposition and cancellation of the allotment granted to the project proponent:

> No one is sure who will clean up the partial construction and people are angry that the buildings are left half-finished and that they dug up sacred land to do this.
>
> (Lokeshwar, Panchayat Pradhan, Chorr project)

Public participation

Public participation (or lack thereof) (30 sources, *67 references*) contained two secondary themes: minimal involvement (26, *36*) and panchayat NOC concerns (17, *31*). The latter reflected that some local residents believed that projects had proceeded without panchayat NOCs being granted (10, *13*), and others believed that clearances had been deceitfully attained (9, *18*) through document forgeries and bribes to village leaders.

Minimal public involvement included the tertiary theme pertaining to lack of opportunities for involvement provided by the project proponent and the state government (15, *19*):

> The State Pollution Control Board does not consult with local communities before making the decision to grant a NOC or not.
>
> (Mohit, state government official)

> They don't consult the persons who are being affected by the small hydro projects. Rather it is a national interest that electricity is to be generated. It is a basic necessity but at the same time the people affected should be consulted.
>
> (Prem, opposition leader, Haripur project)

Minimal public involvement also included the tertiary theme of lack of adequate notice about the project (12, *12*); as one participant noted, "I first heard about the project ten years ago from the sign that was posted at the base of the hill in Bahang village" (Jai Ram, shop owner, Chorr project). At the same time, in some cases, community activists were aware of formal notifications that were given:

> There was press notification that the project was going to be established – is there any objection from the public? So, after that we immediately gave opposition – a letter to the local panchayat. About 300 people signed that letter. So that was given to the panchayat.
>
> (Virbhadra, opposition leader, Haripur project)

Finally, minimal public involvement included the tertiary theme of ways to improve involvement (4, *5*), which captured suggestions from participants for enhancing both early and ongoing community engagement:

> The people of the affected area must be consulted *before* starting the process of the project.
>
> (Prem, opposition leader, Haripur project)

> I'll just say that there should be a committee that should consist of some permanent people of the villages where the project is. There should be representatives from the government. There should be representatives from the company and there should be representatives from Fisheries, Forests, an organization like ours – NGOs which are concerned, really concerned for the public interest, not for their positions.
>
> (Anand, conservation organization)

Community opposition

The fifth primary theme, community opposition (19 sources, *54 references*), included three secondary themes, the most common of which was public protests (13, *23*). Protest actions included petitions, rallies, marches, lobbying and civil disobedience. In the Chorr case, they were especially widespread:

> For the march into Manali, many store owners showed solidarity and closed their shops in protest. Manali mall closes in accordance with the wishes of the president of the Manali market.
>
> (Suman, Panchayat Pradhan, Chorr project)

> They organized a large demonstration in 2011, but the market was closed on three separate occasions for different rallies. At the largest, 30–40 villages were represented, from Kotrang to Solang. Nagar also gathered here.
>
> (Ajay, teacher, Chorr project)

Another secondary theme under community opposition was court challenges (11, *19*), which captured issues, arguments and participant experiences in legal actions brought against three of the projects. Two projects, Chorr and Pakhnoj, ultimately were halted, while the third, Haripur, proceeded and was commissioned in 2017. Community opposition also included a secondary theme covering the roles of community-based organizations (10, *12*), including conservation groups and coalitions of villagers, in opposing the three projects. The tactics of the organizations included forming local committees, accessing government information and maintaining good communication:

> [A local opposition committee formed and] "was successful in part because it recruited an impartial and unofficial leader, whose word was law. He was well educated and a hard worker. He helped to get all the documents."
>
> (Ajay, teacher, Chorr project)

> There is a committee, called the Jan Jagran Avam Vikas Sanstha, which collects information about upcoming projects in this region from government departments or through the Right to Information Act. Once the committee has the information, we all know about the proposed projects.
>
> (Shanta, farmer, Pakhnoj project)

Discussion

The concerns of the research participants about adverse impacts emphasized harm to stream flow regimes, ecological well-being, agriculture and livelihoods. Their views regarding benefits highlighted the potential of the local area development funds, hopes for job creation and economic spinoffs and the belief that small hydro has the capacity to bring benefits to local residents. These results echo findings in similar studies of small hydro in the Kullu District (Kumar and Katoch, 2015; Sinclair, 2003), elsewhere in Himachal Pradesh (Baker, 2014a, 2014b; Rai and Srivastava, 2014), and in the neighbouring Himalayan state of Uttarakhand (Diduck and Sinclair, 2016; McCandless, 2007; Sinclair et al., 2015). A point of departure in this study was the heavy emphasis on concerns about impacts to sacred spaces, reflective of the particular facts of the Chorr case in which such impacts were a flashpoint and became a catalyst for local opposition.

These results suggest the importance of rejecting the rhetoric that small hydro projects have manifold local benefits and minimal adverse impacts. Such statements are prevalent, as seen in the views expressed by project proponents and government officials during this study and in the research cited above. While the negative impacts of small hydro may be less extensive and dramatic than those of large projects, they are no less real and harmful to people, communities and terrestrial systems affected. Minimizing adverse impacts and exaggerating local benefits mask the fact that the projects, in the main, provide electricity to the regional grid in support of industrial activities while harming elements of local social–ecological systems, such as agriculture and traditional livelihoods. These outcomes undermine environmental justice through the inequitable distribution of environmental benefits and risks/ harms (Schlosberg, 2004; Williams and Mawdsley, 2006) and impede the progress of sustainable development by entrenching intragenerational inequities (World Commission on Environment and Development, 1987). They also underscore a need for small hydro planning and approval processes in Himachal Pradesh to more fully attend to the issue of recognitional justice (Schlosberg, 2004; Williams and Mawdsley, 2006). Investigating, understanding and recognizing the diversity of communities, participants, experiences and interests that are affected by proposed projects would improve opportunities for advancing distributive justice when making decisions about small hydro projects.

The results on governance and approvals were descriptive of the roles played by the formally recognized actors (state government agencies and departments, panchayats and project proponents), while the results on public participation were more critical or evaluative. Noteworthy results on public participation were that other than seeking an NOC from affected panchayats, community engagement was non-existent, and provision of notice and public information about the projects were inconsistent. These results are in line with similar studies of small hydro in the Kullu District (Kumar and Katoch, 2015; Sinclair, 2003) and other parts of Himachal Pradesh (Baker, 2014a, 2014b; Rai and Srivastava, 2014). The results about lack of notice and public information also echo concerns raised in studies of large hydro in Himachal Pradesh and Uttarakhand (Diduck, Pratap, Sinclair and Deane, 2013; Diduck, Sinclair, Pratap and Hostetler, 2007; Sinclair and Diduck, 2000) and other industrial projects to which the EIA Notification 2006 was applicable (e.g. Dilay, Diduck and Patel, 2020; Chapters 8 and 9, this volume).

The results on public participation indicate the importance of making small hydro planning and approval processes in Himachal Pradesh more inclusive. The processes found in this study failed to provide opportunities for affected communities to participate in a meaningful fashion by, for example, being involved early and often and having access to complete project information (Diduck and Sinclair, 2021; Stewart and Sinclair, 2007). This shortcoming suggests that Himachal Pradesh is not taking full advantage of opportunities to advance procedural justice (Pring and Pring, 2009) and good governance (United Nations Economic and Social Commission for Asia and the Pacific, 2008) and UN Sustainable Development Goal 16 in respect to promoting inclusive societies and building effective, accountable and inclusive institutions (United Nations Department of Economic and Social Affairs, 2020).

Nevertheless, the results on community opposition reveal successes with respect to access to justice and community-level institutions. The court challenge that blocked the Pakhnoj project reinforces the efficacy of public interest litigation in India under the right conditions (Rajamani, 2007; Chapters 5 and 6, this volume), such as having highly capable legal and technical support and the ability to take on the risks of litigation (Pring and Pring, 2009). Furthermore, the public protests, which had widespread support, that stopped the Chorr project reveal that neoliberal public policies (see the analysis in Chapter 2) may have stifled but not totally eroded norms, customs and organizations in support of local values and aspirations.

Conclusion

The results and discussion lead us to conclude that the state of small hydro development in Himachal Pradesh is sorely wanting and does not support the achievement of, or opportunity for, basic environmental justice for local communities. Local residents revealed the adverse impacts of small hydro, lack of fulfilment of promised benefits and lack of access to local decision processes. These problems led to public protests and legal opposition that came at a cost to the residents and their communities, which could have been avoided with more inclusive and forward-looking small hydro planning and decision-making.

The fundamental basis for environmental justice – recognition of the diversity of participants, experiences and interests is not being acted on – and the inability to easily access NOCs and like processes obstructs procedural justice. To address these shortcomings and thereby enhance the likelihood of distributive justice, small hydro proponents and regulators need to recognize the knowledge and acumen of local mountain people and establish meaningful ways for them to participate in planning and decision-making. Additionally, the findings reveal process concerns that likely cannot be solved through simply tweaking the current planning and decision-making frame. The results underscore the conclusions others have drawn (Diduck and Sinclair, 2016; Erlewein, 2013) that there is a need for strategic assessment of hydro policy, the outcomes of which could help inform DPRs and decision-making for individual projects. We also see the need for regional cumulative effects assessments focused on river catchments in Himachal Pradesh. Such assessments, when done properly, would allow for more thoughtful consideration of impacts of any proposed project in combination with impacts already present (Erikstad, Hagen, Stange and Bakkestuen, 2020). For example, such studies could ensure adequate water availability for irrigation and other purposes. As such, they would help bring to light the full range of environmental, social and economic impacts and their spatial and temporal complexities, along with effective ways to avoid, mitigate or remediate adverse impacts while optimizing benefits.

Inclusive strategic and regional cumulative effects assessments would also open the door to using the wide array of community engagement processes and methods that have proven effective in impact assessment, while shining a brighter light on the impacts of small hydro and especially those that are of great concern to locals and state and national governments, like water availability, climate change and economic development. A promising engagement method that should be considered is the establishment of ongoing community-based management committees, like those established for joint forest management. Such management bodies have proven effective for relationship building and mutual learning. Implementation of such actions would surely advance environmental justice, good governance and sustainable development in Himachal Pradesh. More opportunities to engage in a deliberative way in small hydro decision-making would be possible, and better information on which to base decisions would be available. Inclusive strategic and cumulative effects assessments would also facilitate capacity for more effective integration of small hydro planning into Indian policy positions on such global challenges as climate change and disaster risk reduction. Such integration is consistent with the state's goal to mainstream disaster risk reduction across all government departments (Himachal Pradesh State Disaster Management Authority, 2017), and it would assist in national efforts to attain the broader portfolio of the UN Sustainable Development Goals.

Notes

1 Baloot Fozal 4.6 MW; Baragran 3 MW; Brahim Ganga 5 MW; Fozal 6 MW; Galang 3.5 MW; Hamshu 0.5 MW; Kesta 4.5 MW; Pharari 0.5 MW; Pharari Nallah 0.25 MW; Sarbari-1 4.5 MW; Sarwari-III 2 MW; Sheel 1.5 MW; Shirir 1 MW; Solang 3 MW; Suman Sarwari 5 MW.
2 As noted above and in Chapter 2, many of these NOCs were eliminated from the approval process in 2014.

References

Baker, J.M. (2014a) "Small hydropower development in Himachal Pradesh: An analysis of socioecological effects," *Economic and Political Weekly*, 49(21), 77–86.

Baker, J.M. (2014b) "The socio-ecological effects of small hydropower development in Himachal Pradesh," South Asia Network on Dams, Rivers and People, 8 June. Available at: https://sandrp.wordpress.com/2014/06/08/the-socio-ecological-effects-of-small-hydropower-development-in-himachal-pradesh/ (Accessed: 31 August 2020).

Baker, J.M. (2014c) "The socio-ecological effects of small hydropower development in Himachal Pradesh Part 2," South Asia Network on Dams, Rivers and People, 11 June. Available at: https://sandrp.wordpress.com/2014/06/11/the-socio-ecological-impacts-of-small-hydropower-projects-in-himachal-pradesh-part-2/ (Accessed: 31 August 2020).

Corbin, J. and Strauss, A. (2014) *Basics of Qualitative Research: Techniques and Procedures for Developing Grounded Theory.* 4th edn. Los Angeles, CA: Sage.

Creswell, J.W. (2014) *Research Design: Qualitative, Quantitative, and Mixed Methods Approaches.* 3rd edn. Los Angeles, CA: Sage.

Diduck, A.P., Pratap, D., Sinclair, A.J. and Deane, S. (2013) "Perceptions of impacts, public participation, and learning in the planning, assessment and mitigation of two hydroelectric projects in Uttarakhand, India," *Land Use Policy*, 33, 170–182.

Diduck, A.P. and Sinclair, A.J. (2016) "Small hydro development in the Indian Himalaya: Implications for environmental assessment reform," *Journal of Environmental Assessment Policy and Management*, 18(2), 24 pp.

Diduck, A.P. and Sinclair, A.J. (2021) "A learning-focused analysis of Canada's new Impact Assessment Act" in Doelle, M. and Sinclair, A.J. (eds.), *The New Canadian Impact Assessment Act (IAA).* Toronto: Irwin Law.

Diduck, A.P., Sinclair, A.J., Pratap, D. and Hostetler, G. (2007) "Achieving meaningful public involvement in the environmental assessment of hydro development: Case studies from Chamoli District, Uttarakhand, India," *Impact Assessment and Project Appraisal*, 25(3), 219–231.

Dilay, A., Diduck, A.P. and Patel, K. (2020) "Environmental justice in India: A case study of environmental impact assessment, community engagement and public interest litigation," *Impact Assessment and Project Appraisal*, 38(1), 16–27.

Directorate of Census Operations Himachal Pradesh (2011) *District Census Handbook Kullu.* Shimla, India.

Erikstad, L., Hagen, D., Stange, E. and Bakkestuen, V. (2020) "Evaluating cumulative effects of small scale hydropower development using GIS modelling and representativeness assessments," *Environmental Impact Assessment Review*, 85, 106458.

Erlewein, A. (2013) "Disappearing rivers – The limits of environmental assessment for hydropower in India," *Environmental Impact Assessment Review*, 43, 135–143.

Gill, G.N. (2017) *Environmental Justice in India: The National Green Tribunal.* Abingdon: Routledge.

Government of Himachal Pradesh (2020) *Economic Survey 2019–20.* Shimla: Economic and Statistics Department.

Himachal Pradesh State Disaster Management Authority. (2017). *Himachal Pradesh State Disaster Management Plan.* Shimla: Himachal Pradesh State Disaster Management Authority.

Himurja (2015a) *List/Status of Projects for Which IA's Have Been Signed as on 31/12/2014.* Shimla, India. Available at: http://himurja.nic.in/ongprojects.html (Accessed: 9 March 2015).

Himurja (2015b) *Overview of Small Hydro Programme.* Shimla. Available at: http://himurja. nic.in/smallhydro.html (Accessed: 9 March 2015).

Himurja (2015c) *Small Hydro Development Programme in Himachal Pradesh.* Shimla. Available at: http://himurja.nic.in/invguide.html (Accessed: 9 March 2015).

Johnson, R.M., Edwards, E., Gardner, J.S. and Diduck, A.P. (2018) "Community vulnerability and resilience in disaster risk reduction: An example from Phojal Nalla, Himachal Pradesh, India," *Regional Environmental Change,* 18, 2073–2087.

Johnson, R.M., Edwards, E., Gardner, J.S. and Mohan, B. (2015) "Village heritage and resilience in damaging floods and debris flows, Kullu Valley, Indian Himalaya" in Convery, I., Corsane, G. and Davis, P. (eds.) *Displaced Heritage: Responses to Disaster, Trauma and Loss.* Woodbridge: Boydell Press, 207–224.

Kumar, D. and Katoch, S.S. (2014) "Sustainability indicators for run of the river (RoR) hydropower projects in hydro rich regions of India," *Renewable and Sustainable Energy Reviews,* 35, 101–108.

Kumar, D. and Katoch, S.S. (2015) "Sustainability suspense of small hydropower projects: A study from western Himalayan region of India," *Renewable Energy,* 76, 220–233.

McCandless, M.. (2007) *Community Involvement in the Development of Small Hydro in Uttaranchal, India.* Thesis, Master of Natural Resource Management, University of Manitoba.

Mishra, M.K., Khare, N. and Agrawal, A.B. (2015) "Small hydro power in India: Current status and future perspectives," *Renewable and Sustainable Energy Reviews,* 51, 101–115.

Motupalli, C. (2018) "Intergenerational justice, environmental law, and restorative justice," *Washington Journal of Environmental Law and Policy,* 8(2), 333–361.

Pring, G. and Pring, C. (2009) *Greening Justice: Creating and Improving Environmental Courts and Tribunals.* New Delhi: The Access Initiative.

Rai, A.C. and Srivastava, A. (2014) "Small hydro power projects and community participation" in Singh, R.B. and Hietala, R. (eds.) *Livelihood Security in Northwestern Himalaya: Case Studies from Changing Socio-Economic Environments in Himachal Pradesh, India.* Tokyo: Springer Japan, 239–248.

Rajamani, L. (2007) "Public interest environmental litigation in India: Exploring issues of access, participation, equity, effectiveness and sustainability," *Journal of Environmental Law,* 19(3), 293–391.

Raphael, C. (2019) *Engaged Scholarship for Environmental Justice: A Guide.* Santa Clara, CA: Santa Clara University.

Sah, M.P. and Mazari, R.K. (2007) "An overview of the geoenvironmental status of the Kullu valley, Himachal Pradesh, India," *Journal of Mountain Science,* 4, 3–23.

Schlosberg, D. (2004) "Reconceiving environmental justice: Global movements and political theories," *Environmental Politics,* 13(3), 517–540.

Sharma, N.K., Tiwari, P.K. and Sood, Y.R. (2013) "A comprehensive analysis of strategies, policies and development of hydropower in India: Special emphasis on small hydro power," *Renewable and Sustainable Energy Reviews,* 18, 460–470.

Sinclair, A.J. (2003) "Assessing the impacts of micro-hydro development in the Kullu District, Himachal Pradesh, India," *Mountain Research and Development,* 23(1), 11–13.

Sinclair, A.J. and Diduck, A.P. (2000) "Public involvement in environmental impact assessment: A case study of hydro development in Kullu District, Himachal Pradesh, India," *Impact Assessment and Project Appraisal,* 18(1), 63–75.

Sinclair, A.J., Diduck, A.P. and McCandless, M. (2015) "A comparative case study of small hydro development in the Indian Himalaya" in Dutt, A.K., Noble, A.G., Costa, F.G., Thakur, S.K., Thakur, R. and Sharma, H.S. (eds.) *Spatial Diversity and Dynamics in Resources and Urban Development.* Dordrecht: Springer, 361–378.

Stewart, J.M.P. and Sinclair, A.J. (2007) "Meaningful public participation in environmental assessment: Perspectives from Canadian participants, proponents, and government," *Journal of Environmental Assessment Policy and Management*, 9(2), 161–183.

United Nations Department of Economic and Social Affairs (2020) *Sustainable Development Goal 16: Promote peaceful and inclusive societies for sustainable development, provide access to justice for all and build effective, accountable and inclusive institutions at all levels.* Available at: https://sdgs.un.org/goals/goal16 (Accessed: 29 August 2020).

United Nations Economic and Social Commission for Asia and the Pacific (2008) *What Is Good Governance?* Bangkok: United Nations Economic and Social Commission for Asia and the Pacific. Available at: http://www.unescap.org/pdd/prs/ProjectActivities/Ongoing/gg/governance.asp (Accessed: 29 August 2020).

Williams, G. and Mawdsley, E. (2006) "Postcolonial environmental justice: Government and governance in India," *Geoforum*, 37(5), 660–670.

World Commission on Environment and Development (1987) *Our Common Future.* Oxford: Oxford University Press.

8 A case study of impact assessment, litigation and a social movement against a limestone mine in Gujarat

Ariane Dilay

Introduction

Rising inequalities and environmental degradation in India, resulting from rapid economic growth and industrialization, call for urgent consideration and evaluation of environmental justice. In an effort to enhance procedural justice in the country, the Indian judiciary has established "green benches" in several high courts, as well as a National Green Tribunal (NGT) (Government of India, 2010). Public interest litigation (PIL) and environmental impact assessment (EIA) are also aimed at enhancing justice as they provide opportunities for the public to express their concerns about the industrial developments that have negative social and environmental impacts. Although EIA is designed to help prevent and mitigate negative environmental and social impacts of development, the many challenges and limitations of the process can create conflict between project proponents and affected communities. As demonstrated by this case study, an EIA that is conducted without opportunities for meaningful public involvement can trigger opposition and motivate local communities to seek environmental justice through protests and formation of social movements (Sherpa, Sinclair and Henley, 2015). Such tactics can be successful (Chapters 2 and 3, this volume) but come with serious risks such as legal jeopardy, physical harm and threats to livelihoods (Chapters 9 and 13).

This chapter examines environmental justice within the EIA and environmental clearance process for a controversial limestone mining project in the state of Gujarat. The chapter presents not only a qualitative case study of procedural justice but also touches on recognitional and distributive justice. The case involves public hearings, PIL and a social movement organized in locally affected communities. The case study (1) identifies shortcomings in EIA that induced public opposition and litigation; (2) explains the disconnect between EIA and concerns of local communities, which motivated social movements; and (3) assesses the accessibility of the NGT and other judicial institutions for poor communities seeking environmental justice.

DOI: 10.4324/9781003141228-8

Environmental justice in India

Early approaches to environmental justice focused almost exclusively on fair treatment and equity in the distribution of environmental risk (Hart, 2014; Schlosberg, 2013; Trubek, 1980). Since the 1980s, scholars have expanded this conceptualization to include three distinct attributes: distributive justice aims for equity in the distribution of risk and benefits of development; recognitional justice acknowledges the diversity in participants and experiences in affected communities; and procedural justice focuses on opportunities for participation in environmental policy and decision-making (Hart, 2014; Schlosberg, 2004). A fourth attribute, restorative justice, has been recently included to highlight the need for remediation of negative impacts (Jenkins and Strategist, 2018; Motupalli, 2018). While these four attributes of environmental justice guided this research, the main focus was on procedural justice. In addition, this study was informed by the three pillars of environmental democracy: access to information, access to participation in decision-making and access to justice (United Nations Economic Commission for Europe, 2019).

In 1982, the Supreme Court of India introduced PIL as a tool for individual activists and organizations to seek redress for poor and marginalized sections of society (Bhushan, 2004). Although PIL revolutionized procedural justice in India, many challenges arose as a result of a rapid increase in the number of petitions filed and the intricacy of the scientific cases brought forth (Gill, 2017). In response, the Supreme Court established a bench of environmental judges possessing technical environmental expertise. This type of green bench was similarly adopted in nine state-level high courts. The NGT emerged in 2010 as another mechanism for addressing the inequalities and environmental degradation that had resulted from neoliberal government policies favouring industrial development (Government of India, 2010). These emerging institutional innovations showed a promising effort by the Indian judiciary to advance environmental justice in the country (Gill, 2017).

The quasi-judicial NGT allows any person who is personally, directly or otherwise aggrieved to approach the tribunal and request relief or compensation for victims of environmental damage (Patel and Dey, 2013). The NGT embraces plurality and comprises a multidisciplinary body of both judicial and technical experts (Gill, 2017, 2018). Although the NGT has created a new environmental jurisprudence in India, several scholars have noted its limited scope of authority, a mere seven federal acts, as a shortcoming (Gill, 2017; Nambiar, 2012). The risk of political interference in court decisions, due to the tendency of a neoliberal state to affect quasi-judicial tribunals in India, is also a concern (Dilay, Diduck and Patel, 2020; Gill, 2018). Although these proactive judicial innovations have promoted sustainable development in India and encouraged the use of the precautionary and polluter pays principles, there exist many limitations to their effectiveness (Gill, 2018).

Most natural resources are owned by the state in India, which creates conflicting roles for state governments acting as proprietors, petitioners, regulators and polluters (Patel and Dey, 2013). In response to the often-coercive powers of the state over natural resources, social movements have been on the rise since the 1960s (Desai, 2015; Dilay et al., 2020; Sherpa, Sinclair and Henley, 2015). These movements tend to be made up of socially or politically marginalized people who suffer from distributive injustice, face recognitional or procedural barriers or lack the resources needed to pursue judicial remedies (Bisht and Gerber, 2017; Sherpa, Sinclair and Henley, 2015). As noted earlier, social movements can have positive results (e.g. Chapters 2 and 3) but often come with formidable risks and challenges, including threats of prosecution and physical harm (e.g. Chapters 9 and 13). Understanding the motivation behind social mobilization and how it impacts political decision-making and local community well-being can, thus, shed light on the accessibility – or inaccessibility – of environmental justice in India.

India's EIA system

Chapter 6 of this volume provides a good overview of India's EIA system, and the literature is replete with appraisals of the system's strengths and weaknesses (e.g. Chowdhury, 2014; Dilay et al., 2020; Rathi, 2017). Here I am concerned with selected limitations that have direct implications for environmental justice.

Public engagement has been entrenched in India's EIA system since 1997, when mandatory public hearings were introduced (Agrawal, Lodhi and Panwar, 2010; Diduck, Sinclair, Pratap and Hostetler, 2007; Dilay et al., 2020; Mohan and Pabreja, 2016). In 2006, a new EIA notification process was established in an attempt to increase the efficiency, transparency and objectivity of EIA (Agrawal et al., 2010). Even under the new notification, public participation is only required after the assessment has been conducted, making it extremely difficult to incorporate the public's concerns into planning and decision-making (Agrawal, 2013; Choudhury, 2013; Dilay et al., 2020; Nadeem and Fischer, 2011; Rathi, 2017). Lack of public input early in the planning phase has resulted in lack of trust and rising conflicts between communities and project proponents (Diduck et al., 2007; Diduck, Pratap, Sinclair and Deane, 2013; Mohan and Pabreja, 2016; Nadeem and Fischer, 2011; Paliwal, 2006).

The self-assessment approach and other aspects of EIA in India and elsewhere have been criticized for their inability to ensure environmentally safe and socially acceptable industrial development (Chowdhury, 2014). Other limitations of EIA in India identified in the literature are poor consideration of project alternatives and lack of cumulative impact assessment (Agrawal, 2013; Choudhury, 2013; Diduck et al., 2007; Dilay et al., 2020; Grumbine and Pandit, 2013; Menon and Viswanathan, 2018; Paliwal, 2006; Rathi, 2017).

Research design

Using an exploratory, inductive and adaptive approach (Creswell, 2014; Nelson, 1991), I prepared a qualitative case study focusing on procedural justice within an EIA public hearing, PIL and a related social movement. The primary objective was to obtain the perspectives of individuals in local communities that would potentially be impacted by a limestone mine proposed by Ultratech Cement Ltd. in the Talaja and Mahuva talukas (blocks) of the Bhavnagar District in Gujarat (Figure 8.1). The perceptions of communities that would be directly impacted by the project provide insight into the

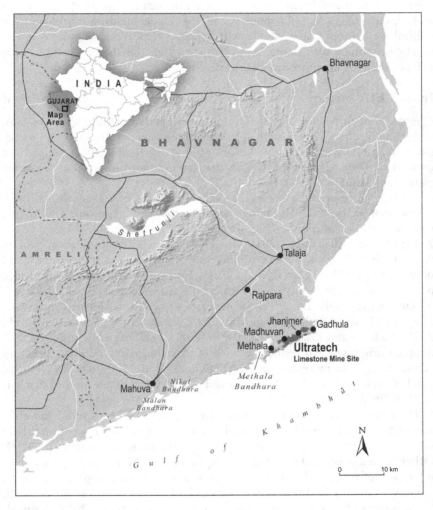

Figure 8.1 Map of the Bhavnagar District, Gujarat, with the location of the proposed Ultratech Cement Ltd. limestone mine.

political legitimacy of the EIA and court system, as well as shed light on the reasoning and motivation for public opposition and social movements. The search for potential case studies began with an extensive review of legal cases while in Canada and after arriving in Gujarat and was followed by consultations with non-governmental organizations and meetings with village leaders in Talaja and Mahuva. The limestone mine was chosen for study because it had potential negative impacts on water resources, and it involved litigation through the NGT and other higher-level courts within the last five years.

Background

Since the advent of liberalization and globalization in India, Gujarat has experienced intensive industrial development and rapid economic growth. The state's economic success is in part attributed to the entrepreneurial nature of Gujaratis and in part to its 16,000 km of limestone-rich coastline (Dholakia, 2007). Limestone extraction for cement manufacturing has become an important activity to meet both local and global demand. However, extensive limestone mining on coastal lands can remove the natural barrier against seawater ingression that the rock provides, causing an increase in salinity in surrounding groundwater and soils (Mahadevia, 1999). This spoilage of freshwater resources in an already water-scarce state can have detrimental impacts to rural communities, the livelihoods of which depend largely on agriculture and the ability to irrigate their lands.

Since 1985, bandharas (a type of tidal regulator) have been constructed in estuaries along the Gujarat coast to help prevent the intrusion of saltwater into freshwater resources and soils (Jadhav and Parasharya, 2004). These bandharas have protected wetland ecosystems that provide habitat for wildlife, including many migratory birds. Bandharas have also allowed surrounding communities to farm for more seasons in the year, thus, reducing the need to migrate for work. In Mahuva taluka, four bandharas have been constructed (Samadhiyala, Malan, Nikol and Kalsar), while conflict was triggered by the construction of a fifth (Methala) (Dilay et al., 2020). Although the Methala bandhara has been approved by the Gujarat government and its construction anticipated by surrounding villages for many years, the Ultratech Cement Company viewed it as a threat to their proposed mining project. Under pressure from the company, the state government put an unofficial hold on the construction of Methala bandhara. As seen in many states in India, local communities are forced to compete with water-intensive industries for access to water and land. These competing interests have become a major driver of public opposition and social mobilization.

Data collection

Using both purposive and convenience sampling to select participants (Barbour, 2001; Creswell, 2014), I conducted semi-structured interviews with stakeholders, including members of rural communities, in proximity to the project site. The purposive sampling sought multiple perspectives and used the criteria of direct involvement in, impact from or knowledge of the case. I attempted to interview project proponents and key government officials but was not granted access to these individuals.

Twenty interviews were conducted from 21 June to 1 August 2017, and lasted between 20 and 75 minutes. An interpreter was present for the interviews, which were audio recorded and later transcribed. Participants were mainly from affected villages and included farmers, agriculture labourers, diamond factory workers and shop owners. Both men and women were interviewed, but most interviewees were men. I also interviewed three provincial politicians, two village political leaders, several village council members, one litigant, one lawyer and one activist, as well as one national and two local non-governmental organizations. I stopped interviewing once data saturation was evident from frequent repetition of themes.

An extensive review of literature and documents was conducted, involving assessments of academic studies, court cases, legislation, media reports, non-governmental organization reports and court-appointed expert committee reports. The EIA report, court filings and public hearing minutes for the Ultratech project were also reviewed.

Participant observation can provide valuable insights into social dynamics and interactions among participants (Vanderstoep and Johnston, 2009). I observed and recorded two village meetings and a foot march to raise awareness of the impacts of the proposed mine through photographs, audio recordings and handwritten notes. Transect walks were also conducted to permit participants to demonstrate the proximity of their land to the proposed mine or to the bandharas, the fertility of their farms and the biodiversity in the area.

Analysis

Thematic analysis of interview responses was conducted using the qualitative data analysis software NVivo (see Braun and Clarke, 2006). Following a grounded theory approach, coding was done inductively, revealing major themes and subthemes in the data (see Tables 8.1–8.3). The tables show the number and percentage of interviews where a theme was mentioned as a way to quantify the data and highlight predominant themes. However, in qualitative research, it is important to recognize that outlier and minority themes are also noteworthy. Direct quotations from interview participants are used to clarify and validate the interpretation of emerging themes.

Documentary, observation and transect walk data were analysed manually and were mainly used to triangulate and validate interview responses. Additionally, observations helped gain a deep understanding of the linkages between culture, politics, livelihoods, natural resources and environment in the study area.

Results

The Ultratech project

Ultratech Cement Ltd. is part of the Aditya Birla Group, India's largest producer and exporter of cement. In 2016, the company drafted a proposal and conducted an EIA for a limestone mine on 1714 ha of land in the Talaja and Mahuva talukas of the Bhavnagar District of Gujarat (J.M. EnviroNet Pvt. Ltd., 2016a,b). The mine is predicted to impact between 70 and 80 thousand people from 13 villages. Three separate public hearings were held by the state pollution control board in June 2016 and these were attended by a total of 10–20 thousand villagers. Primary concerns expressed during public hearings were lack of information about the project, increase in salinity of soil and groundwater and its impact on livelihoods, impacts on plans to construct the Methala bandhara and concerns over land ownership (Public Hearing Proceedings, 2016). Land rights in India are often unclear, and it is common for rural people to remain unaware of their legal rights to land and property (USAID, 2019). In the studied villages, some landowners had previously sold part of their land to another company and/or the government for the construction of the bandhara and were distressed by the idea that their lands could now be acquired by Ultratech Cement Ltd. without their consent. In addition, local people were told that the government is the owner of the minerals underneath the land and, thus, is able to lease the land to Ultratech Cement Ltd. without their approval (Dhar, 2019).

Regardless of the presence of widespread public opposition at all three public hearings, in January 2018, the project received environmental clearance. This was appealed by local activist Bharat Jerambhai Bhil, who filed PIL in the NGT. His primary arguments were based on local knowledge that a forest reserve and mangrove forests were located within a 10 km radius of the project site, critically endangered vultures and migratory birds were found within the project site and there was a risk of disturbance to groundwater and farming activities. The Pune bench of the NGT dismissed the petition against one phase of the project in 2018 (The Indian Express, 2018) and the petition against another phase was dismissed by the Delhi bench of the NGT in February 2019 (DNA, 2019). In September 2019, the Supreme Court ordered the NGT to reopen the case, and a year later the NGT upheld the validity of the environmental clearance (*Gabhabhai Devabhai Chauhan and others v Union of India and others*, 2020). In doing so, the tribunal rejected the arguments of the appellants that the public hearing had been ineffective and the EIA process was flawed for failing to address salinity ingress, adverse impacts on wetlands and cumulative impacts. The NGT expected the local community to provide location-specific scientific evidence that limestone mining would increase salt ingression and affect small-scale agriculture. The community has no knowledge or access to scientific expertise and institutional resources to provide such evidence. With little explanation, on each of these grounds, the tribunal accepted claims made in the EIA report and the positions put forward of the project proponent. The appellants intend to appeal the NGT decision to the Supreme Court (Bharat Bhil, personal communication, 30 October 2020).

Meanwhile, a social movement had formed against the project, and in 2018, mobilized farmers had effectively forced Ultratech to stop project activities on eight separate occasions (The Indian Express, 2018). Demonstrations were still ongoing as of late 2019. The people's movement was led by a former member of the Gujarat legislative assembly (Dr Kanubhai Kalsaria) who also successfully led a people's movement against a Nirma Ltd. cement plant in Mahuva from 2008 until 2015 (Chapter 9, this volume; Dilay et al., 2020). Although the protests began as peaceful marches, violent clashes between police and farmers have been reported and several protesters have been injured and/or arrested (Times of India, 2019). Similar protests have occurred all over India prompted by resentment of local people towards development, often due to inadequate access to information, lack of consultation in early planning phases and disregard of local input in decision-making (Sherpa, Sinclair and Henley, 2015).

Aspects of the EIA process that triggered litigation

Numerous concerns regarding the EIA process were expressed by study participants. These can be divided into two primary themes and two secondary themes (Table 8.1).

Inadequate public participation

In 35 per cent of interviews, participants expressed concerns over not being informed about the project by the proponent. Instead, numerous interviewees explained, they had learned about the project from fellow villagers during public meetings and at similar events. I observed a foot march organized by villagers for the purpose of raising awareness of the potential impacts of the mine. The march was led by Dr Kalsaria and the lead petitioner in the PIL case (Bharat Bhil) and reached thousands of residents throughout five villages surrounding the project site.

Table 8.1 Aspects of EIA that led to court system involvement, as identified by study participants

Primary theme	Secondary theme	Number and per cent of sources (interviews) (N = 20)
Inadequate public participation	Lack of information about the project	7; 35
	Not being heard during the public hearing	11; 55
Inaccurate baseline information		7; 35

In 55 per cent of interviews, participants noted that even though they attended a public hearing and were able to voice their concerns, they were not taken seriously, their concerns were not properly addressed by the project proponent and/or their concerns did not influence decision-making in any way. This was verified to some degree by a review of the public hearing minutes, which revealed that several issues raised by participants were left unaddressed by Ultratech.

Inaccurate baseline information

In 35 per cent of interviews, participants disagreed with the baseline information in the EIA report, mainly with the land classification: Ultratech had identified the project site as wasteland, whereas many people stated that the land was fertile and was used for agriculture, grazing of animals and collecting firewood. For example, one participant noted, "In the EIA report, the company said that this whole area is wasteland. However, we can see different kinds of trees, plants and crops." Several transect walks taken with villagers through their fields made this further evident, as we saw all kinds of crops, fruits and vegetables growing in close proximity to the proposed project site. Some participants suggested that the construction of the Methala bandhara had been purposely delayed to allow Ultratech to classify the land as wasteland: "Without the bandhara, the land looks like a wasteland and, therefore, has a better chance of getting approved for industrial activity, which is good for Ultratech."

The EIA and local concerns

Several concerns over the potential impacts of the Ultratech limestone mine were raised by study participants. Many of these were not addressed in the EIA report and, thus, became key drivers of local opposition. These issues can be divided into two primary and eight secondary themes (Table 8.2).

Table 8.2 Concerns over potential project impacts, as identified by study participants

Primary theme	Secondary theme	Number and per cent of interviews (N = 20)
Environmental impacts	Negative impacts to air and land	7; 35
	Water pollution and increase in salinity	9; 45
	Negative impacts on wildlife	6; 30
Socio-economic impacts	Loss of employment	9; 45
	Increase in seasonal migration	13; 65
	Loss of land	14; 70
	Safety concerns	10; 50
	Health concerns	6; 30

Environmental impacts

In 35 per cent of interviews, participants raised concerns regarding air pollution and an increase in salinity ingress into soils, saying, for example, "If the mine starts there will be lots of dust in the air and the whole land will become a wasteland. Right now, it is fertile because of the limestone; it is a natural layer that prevents the infiltration of sea water into the land." The EIA report fails to comment on the increase in salinity. Similarly, a number of participants (in 45 per cent of interviews) indicated concerns over salinity ingress into water resources. While discussing issues of freshwater shortages, many participants reported the urgent need for the Methala bandhara to be constructed so that the amount of freshwater would increase, allowing for longer farming seasons: "With the bandhara, migration problems will be over and education problems will be over because we will have three seasons of farming."

Concerns regarding the mine's adverse impacts to wildlife in the area were present in 30 per cent of interviews, in statements such as "[I am] worried about living creatures: peacocks, migration birds like flamingos, lions, *nilgais*, leopards and all kinds of trees." Negative impacts to wildlife and to a forest reserve were also mentioned frequently during the public hearings and became the primary argument in the petition filed in the NGT: "Methala village is part of an extension of Gir Forest, where Asiatic lions are observed" (Public Hearing Proceedings, 2016). The company recognized that there is a protected forest and mangrove forests in proximity to the mining site and, thus, proposed to build a green belt to compensate for the loss of biodiversity but claimed that the mangroves had already been degraded by local use and that the mine would not further impact the ecosystem.

Socio-economic impacts

A major issue, raised in 45 per cent of interviews, pertained to the loss of employment if the Ultratech mine was constructed rather than the bandhara. In surrounding villages, construction of the Samadhiyala, Nikol, Malan and Kalsar bandharas has enabled people to farm more productively and for longer seasons. Meanwhile, Ultratech had proposed to hire 30 people from surrounding villages based on their skills and qualifications, but many farmers stated that they did not possess the education or skills needed to work in the mines: "We do not have enough education to be hired, most of us have not graduated grade twelve, most are illiterate and depend on farming." Concerns over loss of employment lead to concerns over seasonal and permanent migration for work (65 per cent of interviews), which were not adequately addressed by Ultratech.

In 70 per cent of interviews, study participants voiced concerns over loss of land: "[The] land here is fertile in natural resources; if the industry extracts the limestone, the land will become wasteland and non-productive." Once again, this local concern was not addressed in the EIA report because the land had been classified as wasteland.

In 50 per cent of interviews, participants expressed worries over the safety of women and children if there were to be an influx of migrant workers. This fear arose from incidents experienced by women in surrounding villages where industries were operating. Health concerns were also mentioned in 30 per cent of interviews, mainly regarding the potential pollution and dust from the mine: "There will be health problems with dust and pollution if Ultratech comes here." Ultratech failed to address the dangers that an influx of outside workers can bring to a community and simply mentioned in their EIA report that the mine would have no significant impacts on human health.

Effectiveness of environmental judicial institutions

Although many participants had limited knowledge of the NGT and other judicial processes, two primary and two secondary themes emerged (Table 8.3).

Accessibility

Although the high cost of accessing the court was only mentioned in 10 per cent of interviews, this was a predominant theme in the literature. Those who did perceive the courts as being costly claimed that lack of financial resources was a major hurdle for poor people in India. One participant noted, "We don't have money but we'll fight with passion." In 25 per cent of interviews, participants expressed the belief that they also lacked the education required to participate in court proceedings. Many felt that their illiteracy and their inability to understand English might result in being misled or misrepresented in a courtroom.

Political influence

A recurrent theme, expressed in 60 per cent of interviews, was that the ruling political party had influence over decision-making in the EIA and environmental clearance processes and favoured industry over the well-being of rural communities. For some, this was made evident by the government's decision

Table 8.3 Participant opinions of environmental judicial institutions

Primary theme	Secondary theme	Number and per cent of interviews (N = 20)
Accessibility	High cost of accessing the courts	2; 10
	Education required to attend and participate in court procedures	5; 25
Political influence	Political influence during the EIA, environmental clearance and in the courts	12; 60

to allow construction of the Ultratech mine and delay construction of the bandhara. In addition, participants expressed concerns that the ruling political party would also be able to interfere in court rulings, that "the central government is completely pro-industry at the moment and can unofficially pressure the court to rule pro-industry."

Discussion

Aspects of the EIA process that triggered litigation

In the Ultratech case, although the EIA was largely compliant with the 2006 EIA notification, lack of recognition of local people and lack of consideration of procedural justice contributed to widespread public opposition. Lack of recognitional justice for poor and marginalized communities in project planning resulting in insufficient consideration of procedural and distributive justice, as seen in this case, is a prominent theme in the literature (Chaudhary, McGregor, Houston and Chettri, 2018; Martin et al., 2016).

From the beginning, inadequate dissemination of project details from the proponent to communities instilled a lack of trust in the Ultratech project. In this case, local villagers found out about the project and its potential impacts from a local activist who was knowledgeable about EIA and ecological processes. It can be assumed that without a local key informant, the surrounding villages would have been kept in the dark about most project details. Neglecting to provide information to locally affected communities has repeatedly been found in earlier studies to be a serious flaw of the EIA process (Diduck et al., 2007, 2013; Dilay et al., 2020; Paliwal, 2006; Rajvanshi, 2003; Rathi, 2017). Overall, the public hearings in this case satisfied the requirement of the law but were not used as a tool for incorporating local knowledge and community interests into project planning. Since local residents felt they had a limited ability to influence decision-making, they resorted to litigation and organized protests to pursue procedural justice.

Inaccurate baseline information and poor quality of EIA reports are common problems in the EIA system in India (Dilay et al., 2020; Paliwal, 2006; Rathi, 2017). In this case, one major concern with the EIA report was the inaccurate land classification. The self-assessment approach could in part be responsible for the poor baseline data, due to the motivation of project proponents to downplay the seriousness of project impacts to expedite the approval process (Menon and Viswanathan, 2018). Lack of input from local communities in determining land use can also lead to misclassification of the land, which can cause widespread opposition by land users, as seen in the controversial Nirma case (Dilay et al., 2020). Failure to acknowledge that the government-approved Methala bandhara was highly anticipated by villagers was another major trigger for litigation and protest. Again, this could have been avoided with proper consultation during project planning, which then could have resulted in more informed site selection.

The EIA report and local concerns

Although this case study focuses on procedural justice and lack thereof, the discrepancies between the EIA report and local concerns cast light on the limited consideration of distributive justice in the proposed limestone mining project. The EIA report stated that 30 people would be hired and that the project would generate a fair amount of direct and indirect employment. However, 30 hires out of the 70–80 thousand affected villagers does not provide an adequate solution to the risk of widespread loss of land and livelihoods.

Villagers in this case were not entirely opposed to development but rather desired the construction of the bandhara to allow for more productive agriculture, followed by the development of agro-processing industries that would enhance their current livelihoods: "Panchyat [a village level elected council for governance] want industries that benefit farmers, such a cotton ginning factory, but first they want the bandhara." Although consideration of project alternatives is a key aspect of EIA regimes in many countries (Pope, Bond, Morrison-Saunders and Retief, 2013), it is rarely included in the terms of reference for EIA in India. This can lead to inadequate site location for projects, which is a major impediment to achieving distributive justice (Dilay et al., 2020; Rathi, 2017). Better consideration of recognitional and procedural justice, such as involving communities in early phases of project planning, could have facilitated the realization of community aspirations and goals in this case rather than forcing an unwanted industry in an inappropriate location.

Effectiveness of environmental judicial institutions

Procedural aspects of environmental justice in India have been significantly improved by the judicial innovations of PIL, green benches and the NGT. However, local people still face many challenges in accessing the courts. Limited financial and technical resources create obstacles in approaching the judiciary (Dilay, Diduck and Patel, 2020; Gill, 2018). Villagers in this case were fortunate to have a local activist who had knowledge of the EIA process and who was able to file a PIL on their behalf. This is not always the case in India, as poor citizens are often victims of environmental degradation and can be forced to leave their homes and livelihoods by neoliberal forces pushing for industrialization and economic growth at all costs (see Thompson, 2008).

Since the NGT has dismissed the most recent appeal and has not issued an interim order staying the project, local people are again turning to the Supreme Court, which is a costly endeavour in terms of both time and finances. They have also turned to social mobilization in the form of rallies and protests to prevent construction of the mine. Despite the flaws in the public hearing and the EIA report noted above, the NGT gave little or no credence to the arguments of the communities. With little explanation, it accepted the arguments of the project proponents and by extension the validity of the EIA process and accuracy of the EIA documents.

Despite the judiciary's effort to increase environmental justice in India, social movements are still required by local citizens seeking to preserve their access to natural resources and ecosystem services. Since the 1960s, this form of environmental activism by the poor has been undertaken against mining activities in several states in India (Bisht and Gerber, 2017). Previous studies have suggested that protesters are often motivated by the perception that government favours mining operations over local livelihoods, a sentiment directly expressed by our study participants: "The government is not on our side. There is the possibility that some politicians are business partners with the company while some will take bribes from industry." While local protests in Bhavnagar have had success in preventing Ultratech from mining as of late 2020, this has come at a cost to organizers and participants in the form of potential legal jeopardy, physical harm and threats to livelihoods. A court decision in favour of local people would eliminate such costs and definitively stop the company from mining in the area.

Conclusion

The proposal for the Ultratech limestone mine in the coastal villages of Bhavnagar, Gujarat, followed a common trajectory of inadequate public engagement throughout the project planning, EIA and environmental clearance processes, resulting in public opposition, litigation and protest. This trajectory has become a trend in India, owing to the lack of proactive participation opportunities in the early phases of development. In this case study, procedural aspects of environmental justice were present but found to be weak, while consideration of recognitional and distributive forms of justice were seemingly absent.

Fundamental changes are needed to improve the EIA process and achieve more sound development. Nadeem and Fischer (2011) suggest mandatory public engagement prior to site selection and before conducting the EIA for new projects, which could help align development goals and community aspirations. Establishing more opportunities for information sharing between project proponents and local communities, involving all stakeholders in early planning phases and incorporating local knowledge into project design would be significant steps in achieving recognitional, procedural and distributive justice (Dilay et al., 2020; Momtaz and Gladstone, 2008; Rajvanshi, 2003).

Despite the Indian judiciary's effort to render procedural environmental justice more achievable for marginalized sectors of society, there remain significant challenges to accessing the court system in India. Lack of financial and technical resources to help communities approach the courts could be overcome with the implementation of better participant assistance programs. Since it is often perceived that government, courts and industries are intertwined, more transparency within court processes could help the courts gain public trust.

Aiming to improve recognitional, procedural and distributive justice in India can help to reduce the need for communities to resort to protests and social movements as platforms for pursuing environmental justice. Although such opposition efforts sometimes bear fruit, organizers and participants often pay a heavy price. By providing more frequent and more meaningful opportunities to involve communities in development planning and decision-making, recognitional, procedural and distributive aspects of environmental justice can be improved, which can help local communities reach their goals as well as achieving broader development objectives.

References

Agrawal, D.K., Lodhi, M.S. and Panwar, S. (2010) "Are EIA studies sufficient for projected hydropower development in the India Himalayan region?" *Current Science*, 98(2), 154–161.

Agrawal, R. (2013) "Hydropower projects in Uttarakhand: Displacing people and destroying lives," *Economic and Political Weekly*, 11.8(29), 14–16.

Barbour, R.S. (2001) "Checklists for improving rigour in qualitative research: A case of the tail wagging the dog?" *British Medical Journal (Clinical Research ed.)*, 322(7294), 1115–1117.

Bhushan, P. (2004) "Supreme court and PIL: Changing perspectives under liberalization," *Economic and Political Weekly*, 29(18), 1770–1774.

Bisht, A. and Gerber, J. (2017) "Ecological distribution conflict (EDCs) over mineral extraction in India: An overview," *The Extractive Industries and Society: An International Journal*, 4(3), 528–563.

Braun, V. and Clarke, V. (2006) "Using thematic analysis in psychology," *Qualitative Research in Psychology*, 3(2), 77–101.

Chaudhary, S., McGregor, A., Houston, D. and Chettri, N. (2018) "Environmental justice and ecosystem services: A disaggregated analysis of community access to forest benefits in Nepal," *Ecosystem Services*, 29(Part A), 99–115.

Choudhury, N. (2013) "Legality and legitimacy of public involvement in infrastructure planning: Observations from hydropower projects in India," *Journal of Environmental Planning and Management*, 57, 297–315.

Chowdhury, N. (2014) "Environment impact assessment in India: Reviewing two decades of jurisprudence," *IUCN E Law Journal*. Available at: https://www.researchgate.net/publication/263273040_Environment_Impact_Assessment_in_India_Reviewing_Two_Decades_of_Jurisprudence (Accessed: 1 October 2020).

Creswell, J.W. (2014) *Research Design: Qualitative, Quantitative, and Mixed Methods Approaches*. 4th edn. Los Angeles, CA: Sage.

Desai, M. (2015) *Subaltern Movements in India: Gendered Geographies of Struggle against Neoliberal Development*. London: Routledge.

Dhar, D. (2019) "Farmers protest against land acquisition from limestone mining by Ultratech Cement in Bhavnagar," *NewsClick*, 2 March. Available at: https://www.newsclick.in/farmers-protest-land-acquisition-limestone-mining-ultra-tech-cement-bhavnagar (Accessed: 11 September 2019).

Dholakia, R.H. (2007) "Sources of economic growth and acceleration in Gujarat," *Economic and Political Weekly*, 42(9), 770–778.

Diduck, A.P., Pratap, D., Sinclair, A.J. and Deane, S. (2013) "Perceptions of impacts, public participation, and learning in the planning, assessment and mitigation of two hydroelectric projects in Uttarakhand, India," *Land Use Policy*, 33, 170–182.

Diduck, A.P., Sinclair, A.J., Pratap, D. and Hostetler, G. (2007) "Achieving meaning-ful public participation in the environmental assessment of hydro development: Case studies from the Chamoli District, Uttarakhand, India," *Impact Assessment and Project Appraisal*, 25(3), 219–231.

Dilay, A., Diduck, A.P. and Patel, K. (2020) "Environmental justice in India: A case study of environmental impact assessment, community engagement and public interest litigation," *Impact Assessment and Project Appraisal*, 38(1), 16–27.

DNA (2019) "Supreme Court relief for farmers battling mining by Ultra Tech," *DNA*, 3 September. Available at: https://www.dnaindia.com/ahmedabad/report-supreme-court-relief-for-farmers-battling-limestone-mining-by-ultra-tech-2787001 (Accessed: 11 September 2019).

Gabhabhai Devabhai Chauhan and Others v Union of India and Others, Appeal No. 62 of 2018, order dated 24 September 2020 (National Green Tribunal).

Gill, G.N. (2017) *Environmental Justice in India: The National Green Tribunal*. Oxford: Routledge.

Gill, G.N. (2018) "Mapping the power struggles of the National Green Tribunal of India: The rise and fall?" *Asian Journal of Law and Society*, 7(1). Available at: doi:10.1017/als.2018.28 (Accessed: 1 October 2020).

Government of India (2010) *Report of the Expert Group to Review the Methodology for Measurement of Poverty*. New Delhi: Planning Commission of India, Government of India.

Grumbine, R.E. and Pandit, M.K. (2013) "Threats from India's Himalaya dams," *Science*, 339(6115), 36–37.

Hart, C.R. (2014) "The role of environmental justice in biodiversity conservation: Investigating experiences of communities near Kruger National Park, South Africa," *Dalhousie Journal of Interdisciplinary Management*, 10(1). Available at: https://doi.org/10.5931/djim.v10i1.3359 (Accessed: 4 April 2019).

Jadhav, A. and Parasharya, B.M. (2004) "Counts of flamingos at some sites in Gujarat state, India," *Waterbirds*, 27(2), 141–146.

Jenkins, B. and Strategist, S. (2018) "Environmental restorative justice: Canterbury cases" in *Environmental Justice in Societies in Transition: Proceedings of the 38th Annual Conference of the International Association for Impact Assessment*. Adelaide, Australia. Available from: iaia.org [11 May 2019].

J.M. EnviroNet Pvt. Ltd. (2016a) *Final Environmental Impact Assessment Report and Environmental Management Plan for Proposed Limestone Mine (M.L. Area-193.3269 ha) with Production Capacity of 3.21 MTPA (ROM) at Villages-Methala, Madhuvan, Jhanjmer, Rajpara, Ralia and Gadhula, Taluka-Talaja, District-Bhavnagar (Gujarat)*. Available at: https://www.google.com/url?sa=t&rct=j&q=&esrc=s&source=web&cd=&ved=2ahUKEwj2op3Lza_sAhUDa80KHaw8D6sQFjAAegQIBRAC&url=http%3A%2F%2Fenvironmentclearance.nic.in%2Fwritereaddata%2FEIA%2F120820160QVO7F-4HFINALEIAEMPREPORT.pdf&usg=AOvVaw3S1hsTeuSPPBHOYibukEJz (Accessed: 5 May 2019).

J.M. EnviroNet Pvt. Ltd. (2016b) *Final Environmental Impact Assessment Report and Environmental Management Plan for Proposed Limestone Mine (M.L. Area-851.3216 Ha) with Production Capacity of 0.53 MTPA at Villages-Bambhor and Talli, Taluka-Talaja, District-Bhavnagar (Gujarat)*. Available at: https://www.google.com/url?sa=t&rct=j&q=&esrc=s&source=web&cd=&ved=2ahUKEwi00Indzq_sAhVXbs0KH-Q2vDbAQFjABegQIAxAC&url=http%3A%2F%2Fenvironmentclearance.nic.in%2Fwritereaddata%2FEIA%2F16082016Y9A0Q0Q4FinalEIA.pdf&usg=AOvVaw2BsKvyy9-oRQdakik2FEhQ (Accessed: 5 May 2019).

Mahadevia, D. (1999) *Economic Growth and Environmental Degradation: Case of Gujarat.* New Delhi: Research Foundation for Science, Technology and Ecology, 1–165.

Martin, A., Coolsaet, B., Corbera, E., Dawson, N.M., Fraser, J.A., Lehmann, I. and Rodriguez, I. (2016) "Justice and conservation: The need to incorporate recognition," *Biological Conservation*, 197, 254–261.

Menon, M. and Viswanathan, V. (2018) "How not to do an environmental assessment," *The Hindu*, 30 August. Available at: https://www.thehindu.com/opinion/op-ed/how-not-to-do-an-environmental-assessment/article24813642.ece (Accessed: 30 August 2019).

Mohan, M. and Pabreja, H. (2016) "Public hearings in environmental clearance process: Review of judicial intervention," *Economic and Political Weekly*, 11(50), 68–75.

Momtaz, S. and Gladstone, W. (2008) "Ban on commercial fishing in the estuarine waters of New South Wales, Australia: Community consultation and social impacts," *Environmental Impact Assessment*, 28(2–3), 214–225.

Motupalli, C. (2018) "Intergenerational justice, environmental law, and restorative justice," *Washington Journal of Environmental Law and Policy*, 8(2), 333–361.

Nadeem, O. and Fischer, T.B. (2011) "An evaluation framework for effective public participation in EIA in Pakistan," *Environmental Impact Assessment Review*, 31, 36–47.

Nambiar, S. (2012) "Paradigm of 'green' adjudication: Developing principles for Indian environmental decision-making in disputes involving scientific uncertainty," *ILI Law Review*, 1(1), 1–24.

Nelson, J.G. (1991) "Research in human ecology and planning: An interactive, adaptive approach," *Canadian Geographer*, 35(2), 114–127.

Paliwal, R. (2006) "EIA practice in India and its evaluation using SWOT analysis," *Environmental Impact Assessment Review*, 26(5), 492–510.

Patel, K. and Dey, K. (2013) "The trajectory of environmental justice in India: Prospects and challenges for the National Green Tribunal" in Tim, M., Trivedi, N. and Vajpeyi, D. (eds.), *Perspectives on Governance and Society*. New Delhi: Rawat Publications, 160–174.

Pope, J., Bond, A., Morrison-Saunders, A. and Retief, F. (2013) "Advancing the theory and practice of impact assessment: Setting the research agenda," *Environmental Impact Assessment Review*, 41, 1–9.

Public Hearing Proceedings (2016) *Annexure-B (ENGLISH),* Issues raised by affected people and replies given by the representative of the project proponents. Available at: https://www.google.ca/url?sa=t&rct=j&q=&esrc=s&source=web&cd=&ved=2ahUKEwju-5dai1K_sAhWRPM0KHexLD60QFjAAegQIAhAC&url=https%3A%2F%2Fgpcb.gujarat.gov.in%2Fhearingpdf%2FULTRATECH_CEMENT_LTD_BHV28_PH_MOM3_PRO.PDF&usg=AOvVaw3mV75kda5EuGyHH2ORLPEO (Accessed: 30 May 2019).

Rajvanshi, A. (2003) "Promoting public participation for integrating sustainability issues in environmental decision-making: The India experience," *Journal of Environmental Assessment Policy and Management*, 5(3), 295–319.

Rathi, A.K.A. (2017) "Evaluation of project-level environmental impact assessment and SWOT analysis of EIA process in India," *Environmental Impact Assessment Review*, 67, 31–39.

Schlosberg, D. (2004) "Reconceiving environmental justice: Global movements and political theories," *Environmental Politics*, 13(3), 517–540.

Schlosberg, D. (2013) "Theorising environmental justice: The expanding sphere of a discourse," *Environmental Politics*, 22(1), 37–55.

Sherpa, Y.D., Sinclair, A.J. and Henley, T. (2015) "'Stepping on the heads of our gods': Community action and learning in response to tourism development in Manali, India," *International Journal of Social Ecology and Sustainable Development*, 6(2), 40–56.

The Indian Express (2018) "Bhavnagar District: Farmers force firm to suspend mining," *Express News Service*, 27 December. Available at: https://indianexpress.com/article/cities/rajkot/bhavnagar-district-farmers-force-firm-to-suspend-mining-5511137/ (Accessed: 20 September 2019).

Thompson, M. (2008) *Organizing and Disorganizing: A Dynamic and Non-Linear Theory of Institutional Emergence and Its Implications*. Devon: Triarchy Press.

Times of India (2019) "Rally against mining turns violent near Talaja, 15 hurt," *Times of India*, 3 January. Available at: https://timesofindia.indiatimes.com/city/rajkot/rally-against-mining-turns-violent-near-talaja-15-hurt/articleshow/67355653.cms (Accessed: 20 September 2020).

Trubek, D.M. (1980) "Studying courts in context," *Law and Society Review*, 15(3–4), 485–502.

United Nations Economic Commission for Europe (2019) *Convention on Access to Information, Public Participation in Decision-Making and Access to Justice in Environmental Matters*. United Nations Economic Commission for Europe. Available at: https://www.unece.org/env/pp/contentofaarhus.html (Accessed: 14 January 2019).

USAID (2019) "India: New country profile," *LandLinks*. Available at: https://www.land-links.org/country-profile/india/#1529353387413-8c48e472-f4d0 (Accessed: 14 January 2019).

Vanderstoep, S.W. and Johnston, D.D. (2009) *Research Methods for Everyday Life: Blending Qualitative and Quantitative*. San Francisco, CA: Wiley.

9 Values matter

Gender-based exclusion from environmental impact assessments in Mahuva, Gujarat

Bryce Gallant

Introduction

Environmental impact assessments (EIAs) have become an important tool shaping environmental management (Morgan, 2012) and for achieving environmental justice. Designed to avoid or mitigate potential adverse changes to the environment (including biophysical, social and economic impacts) caused by major development projects such as dams, mines and industrial facilities, an EIA predicts potential impacts and outcomes. Thus, the values used in EIAs play a crucial role in determining the environment's worth, as they influence such aspects as which voices are heard, what knowledge is considered, which biophysical components are considered, how the environment's monetary value is determined and what protective measures are designated. Part of the assessment process usually requires a public hearing in affected communities to incorporate local voices and concerns (Sainath and Rajan, 2015). This can give the hearing panel and project proponents the opportunity to learn from local knowledge and address concerns that are raised. This makes the EIA process an important mechanism for community members to formally participate in environmental management.

When utilized well, EIAs can help strengthen recognitional and procedural justice. Authentic recognition of individuals, communities, cultures and collective identities helps mitigate issues of disenfranchisement, through respect for and acknowledgement of diverse sovereign voices (Schlosberg, 2004). Related to this, procedural justice is concerned with the proper use of power: the nature of the power used, by whom, over whom and to what ends (Howe et al., 2012). Procedural justice requires ensuring different groups are actively involved in processes like EIAs to enable meaningful engagement with formal environmental justice mechanisms. Facilitating recognition, participation and increased inclusion requires a shift in how EIAs currently value the environment – to a process that incorporates a variety of perspectives, knowledge and understandings that fully represent local experiences and diversity.

DOI: 10.4324/9781003141228-9

To ensure communities and their heterogenous nature are captured well, gender-based analysis (GBA) is being used in some EIA systems. For example, Canada has now made its inclusion a legal requirement (Government of Canada, 2019; Impact Assessment Agency of Canada, 2019), while the increasing importance of GBA is also being recognized in Europe and by organizations such as World Bank. However, GBA is not being implemented everywhere or equally, with many governments struggling to incorporate GBA in EIAs. Ignoring gendered understandings of the environment can result in EIAs failing to consider important values such as non-monetary uses and connections to land that are critical to communities' ways of life.

Using a case study of cement company Nirma Ltd. (referred to as Nirma Cement) in Mahuva taluka (block), Gujarat, this chapter argues that EIAs are a gendered process and focus on particular kinds of values and knowledge that often exclude experiential perspectives and knowledge – prompting community mobilization. The underpinning values and normative assumptions, in combination with faulty community participation mechanisms (including formal public hearings), often create barriers that push women out of the formal EIA process – removing their ability to use this important environmental management tool.

The following section briefly examines limitations of EIAs and how knowledge and values shape EIA practice. The next section looks at valuation systems and the use of informal spaces in mobilization, which is followed by the research methodology and case study. The discussion examines the types of valuation systems present, how values and knowledge are legitimized in EIAs and how exclusion from formal environmental justice mechanisms can lead to collective mobilization. Finally, the conclusion emphasizes the need to change current EIA valuation systems to be more inclusive of marginalized knowledge.

How knowledge and values shape EIA practice

Scholars have scrutinized EIA legislation and practice in India (e.g., Paliwal, 2006; Rathi, 2017). Limitations of EIAs include low quality reports, lack of consideration of alternatives, poor public engagement processes, lack of gender as a unit of analysis and deficient monitoring mechanisms (Diduck et al., 2013; Erlewein, 2013; Morgan, 2012; Rathi, 2017). EIAs are also based on implicit knowledge (Slootweg, Vanclay and van Slootweg, 2003) and too often on the assumption of being gender neutral. Yet environmental impacts have multiple and different consequences for men, women and children (Srinivasan and Mehta, 2003). What those conducting EIAs value about the environment influences which aspects are considered in official EIA reports and what the public engagement process looks like (Wilkins, 2003).

Research has also found that consultation and engagement activities often focus on concerns and impacts associated with typically male-dominated activities, while those usually associated with women are marginalized or omitted (Mills, Dowsley and Cameron, 2014). Without analysing power, the embedded social and cultural beliefs that disadvantage women (Joshi, 2011) go unresolved and are reproduced. To disregard gender is to ignore the basis of social, economic, cultural and political organization of a society or community (Srinivasan and Mehta, 2003) and usually results in marginalized groups being shut out of formal environmental justice processes. Therefore, not paying attention to existing social contexts and power relationships before new resource developments are introduced may simply exacerbate pre-existing inequalities rather than reduce them (Hill, Madden and Collins, 2017). Thus, the chosen focus of an EIA shapes who and what knowledges are included, whose impacts are considered and mitigated and the processes by which decisions are made (Walker and Reed, 2021).

Systems of environmental valuation

Ecological, political and relational dimensions vary across contexts, shaping the relationships people have with their environment. Just as environmental knowledge is based in systems characterized by power discrepancies and unequal social relations (Rocheleau, Thomas-Slayter and Wangari, 1996), so are the valuation systems upon which this knowledge is based. This is why in EIAs, generalizations tend to lead to exploitation, misunderstandings, power inequalities and concerns about management (Perkins, 2007).

The processes of valuing and devaluing nature, knowledge and people shape the ways in which harms and benefits are distributed (Burke and Heynen, 2014). In EIAs, this process often involves subjective decision-making about environmental components (Mostert, 1996). Three common systems of environmental valuation have been identified: corporate-oriented, government-oriented and household-oriented (Burke and Heynen, 2014). Corporate- and government-oriented systems emphasize lucrative exchange values that can be uncounted or devalued based on who is going to receive benefits, while being influenced by property ownership, control, access and regulation (Ayyub, 2014). In other words, cost-benefit analyses are used to quantify the environment (EPA (United States Environmental Protection Agency), 2014) and derive value from its usefulness in achieving a goal (Kula and Evans, 2011; Ayyub, 2014). The household-oriented system of valuation relies on personal knowledge of local environments that has been collected through experiences and practices and the economic activities that comprise a household's livelihood (Burke and Heynen, 2014; Pascua et al., 2017). This system acknowledges and values that local communities possess extensive knowledge of relationships among terrain, climate, vegetation and animal nutrition and behaviour (Fernández-Giménez and Estaque, 2012). Each system reinforces a particular way of knowing nature.

Knowledge is validated and legitimized in these value systems. And power is exercised through environmental knowledge and the discourses that legitimize dominant knowledge claims (Bassett and Peimer, 2015; Forsyth and

Walker, 2008). However, the point here is not about leveraging one under-standing of the environment over the other but instead creating a deeper, more complex understanding built on a variety of perspectives and values (Fernández-Giménez and Estaque, 2012; Rice et al., 2015).

The importance of informal spaces

Gender and social norms shape the values that contribute to women's exclu-sion from various environmental processes, such as EIAs (Agarwal, 1998). Limited access to formal environmental justice mechanisms can result in informal spaces facilitating mass mobilization, a shift out of constrained spaces and widening the scope for action (Bulota, 2018; Cornwall, 2002).

When disadvantaged peoples have no voice within a group or system, they may opt to set up their own group and work outside the system. For example, protests can be more visible and inclusive alternatives than officially sanc-tioned decision processes (Mavroudi, 2008). Protests become a space where community members can come together to create solidarity and enable an exchange of ideas, knowledge, practices, materials and resources (McFarlane, 2009). This often gives community members a space where they feel more in control, allowing them to represent themselves and their environmental knowledge, and get their message of protest across to a wider audience as a collective (Mavroudi, 2008).

However, the rules, norms and practices that shape society and institu-tions may be reiterated within participatory, informal spaces. This can result in participatory exclusions, which Agarwal (2001) describes as "exclusions within seemingly participatory institutions" (p. 1623), which often affect marginalized groups such as women. In other words, informal spaces are not gender neutral. The case study below will examine how a people's move-ment (andolan) was used to engage in and influence the environmental justice process, the gendered nature of the andolan and the trade-offs involved in participating in the movement.

Research design

An exploratory and inductive approach was taken for a qualitative case study of an EIA, community engagement and collective mobilization. The case focuses on rural women's participation in the EIA process, how they value their physical and social environments and the conditions that resulted in women utilizing informal spaces. Local perspectives were used to understand and contextualize the motivations that led to the opposition against Nirma Cement. In particular, informal interactions with community leaders were crucial in gaining access to communities affected by Nirma Cement.

The primary method of data collection was semi-structured interviews. In total, 22 interviews – a mixture of individual and small group discus-sions, both formal and informal – were conducted with local stakeholders from 21 June to 1 August 2017. Interviews, which involved an interpreter, lasted between 20 and 75 minutes; audio was recorded and later transcribed. Both men and women were consulted. Most participants were villagers and included farmers, agricultural labourers, diamond factory workers, shop owners, small-scale factory owners, doctors and tailors. The participants also included three provincial politicians, two village political leaders, numerous village council members, one litigant, one lawyer and one activist, as well as representatives of one national and two local non-governmental organizations. Efforts were made to interview the project proponent and key government officials to gain their perspectives, but they did not provide access.

Participation observation was also used, which allowed a better understanding of group dynamics and social relations. Additionally, transect walks enabled participants to demonstrate the fertility of their soils, the types and number of livestock they owned, the proximity of their land to a freshwater source and the partially built Nirma Cement infrastructure and the biodiversity in the region, as well as communicate their values about each aspect. This data was recorded with handwritten notes, photographs, maps and diagrams.

Building on preliminary analysis that was done in the field, NVivo 11 was used to perform an inductive thematic analysis of the interview responses (Braun and Clarke, 2006). The observational data was used to supplement or validate the interview responses and gain understanding of local culture, gender power dynamics, environments and livelihoods. Women's testimonials were combined with an extensive literature and document review, examining academic studies, court filings and decisions, legislation, investigative media reports, non-governmental organization reports, court-appointed expert committee reports and EIA documentation.

Case study: Nirma Cement, Mahuva, Gujarat

Mahuva taluka is an area located on the Arabian Sea coast of the Bhavnagar District in Gujarat, India, known for its vast amount of limestone. Figure 9.1 shows the case study area.

Being a coastal area, Mahuva has several river tributaries and streams that flow towards the sea and provide water for small-scale agriculture in the region. However, these started drying due to a series of dams and over-exploitation of water upstream over the last two to three decades. Farming and livestock rearing suffered severely because of the water shortage, which forced many people to migrate during summer months. The tidal water from the sea also started flowing inland into dried estuaries and rivers, which made the soil and the groundwater saline in the region, and this salt ingression has long-term impacts on soil and farming. This resulted in many farmers in the Mahuva block struggling to produce crops (Sheth and Raval, 2014). In early 2000, however, the government of Gujarat launched a scheme to build band-haras, 1.8–3 m tall dykes or dam-like structures at the mouths of rivers or estuaries before they merge with the sea. Bandharas create a small reservoir

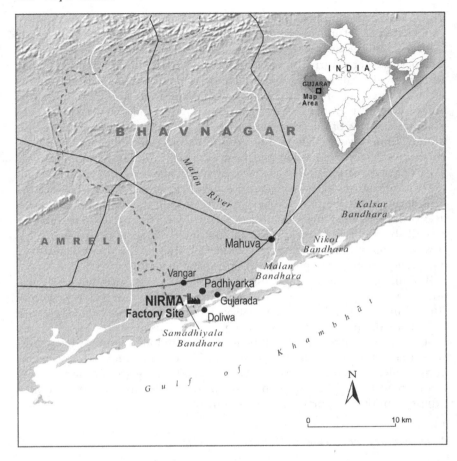

Figure 9.1 Map of the Bhavnagar District, Gujarat, with the location of the proposed Nirma Cement factory.

of freshwater, which farmers can access for a variety of purposes. They also prevent tidal water from flowing into the river, hence reducing salt ingression of surface and groundwater. These bandharas, thus, support livestock and have allowed farmers to irrigate their crops throughout the year.

Women emphasized the positive impact bandharas have had on their livelihoods by enabling easier access to drinking water for animals, increased vegetable farming and recharging of subsurface water, which is drawn for their consumption using handpumps. For them, a bandhara is the only thing that reduces the risk of rural outmigration due to failure of agriculture and loss of livelihoods. They highlighted how quality of life and well-being have improved in the villages where the government has built bandharas in the region. For example, women indicated that they prefer their daughters to marry in villages that have access to bandhara water so they do not have to depend on the income of their husband from either migrant labour work or a job in a local diamond factory.

In 2008, the state government allotted part of the Samadhiyala bandhara and its catchment area to Nirma Cement for a cement plant, coke oven and power plant (Dilay, Diduck and Patel, 2019). Permission was also granted for mining on private land, which would remove limestone and encourage subsurface flow and salt ingression, making fertile agriculture soil saline. The EIA for the project was then conducted in 2009. The assessment examined an area with a 10-km radius for possible impacts on air, soil, water, ecology, land use patterns and socio-economic implications. The report identified five villages that could be affected: Dhugery, Doliya, Gujarada, Mahuva and Vangar. It found a variety of major environmental issues that could arise: air pollution, loss of biodiversity, desertification, drought, food insecurity, soil contamination, soil erosion, deforestation and loss of vegetation, water pollution, large-scale disturbance of hydro and geological systems and reduced ecological connectivity (MIN MEC Consultancy Pvt. Ltd., 2008). The EIA report also identified the study area's population of 134,069, general demographics, literacy rates and amenities. It claimed the project would provide a variety of benefits, including zero displacement, direct employment of 418 people and additional indirect employment, alternative arrangements of livelihoods for those who lost land, a supply of drinking water and development of facilities within villages (MIN MEC Consultancy Pvt. Ltd., 2008).

What the EIA failed to mention is that the area allocated for construction is a common property wetland used for a variety of purposes, including animal grazing, firewood collection, irrigation and drinking water. It also failed to identify social issues that community members had highlighted, such as displacement, job insecurity, higher unemployment rates, loss of livelihood, loss of traditional knowledge/practices/cultures, land dispossession and a loss of landscape and sense of place (Dilay, Diduck and Patel, 2019; Sheth and Raval, 2014; Shrivastava, 2015)

In Mahuva, land matters to the community's livelihoods and overall way of life. Most people in the area rely on farming as their main source of income and food. An increase in water salinity could significantly impact crop production of the community members. This decline in production, combined with reduced land for animal grazing, could result in widespread food insecurity. It could also lead to mass migration, contributing additional financial and emotional stress as many families cannot afford to relocate or have nowhere else to go.

Emergence of the Mahuva Andolan

The project proponents and government officials held the public hearing for the Nirma Cement EIA in 2008. However, many villagers noted that they had been unaware of the EIA report before the hearings, resulting in a lack of awareness of and participation in the process. Records indicate that 50 members of the public attended the hearing, with 13 members giving statements (Gujarat Pollution Control Board, 2008), despite thousands of people being affected. No women were mentioned as being present in the official public hearing participant lists.

Despite the official report, many women did attend. A prominent woman in the area gathered a large group of community members, mostly women, to speak at the hearing. The chairman of the hearing made it clear that everyone in attendance had a right to speak and invited these women to share their concerns. However, when they went to speak, the panel members walked out, resulting in the chairman shutting down the meeting early. Angered that their voices were being ignored, the women gathered at the Nirma Cement project site and started to dismantle what construction had been started. Nirma Cement retaliated by lodging police complaints against those who participated, with some spending time in jail. This incident sparked the creation of the Mahuva Andolan, or Mahuva Protest – an informal space where voices that had been marginalized in the formal process, could oppose Nirma Cement's plans.

Since that initial incident, there have been a variety of marches and rallies under the umbrella of the Mahuva Andolan. One of the most famous protests was a march from Doliya to Gandhinagar, on 3–17 March 2011, which spanned 170 km. On this occasion the women in the group made food items out of cement to present to political leaders as a symbolic gesture to emphasize the importance of farming for local communities. Additional protests included marches between villages in affected areas and public rallies and meetings.

The women who started the andolan did so because for them the local environment is not just a place to live and profit from but represents a way of life. It connects people to their histories and provides a sense of place. As the andolan evolved, the women who founded it have remained active participants and planners in protests, while local political figures have taken on leadership positions to increase awareness and build on gained momentum. To date, over 45,000 local people have been part of the movement and the Nirma Cement factories have never been completed, although the initial construction has remained.

Discussion

Environmental valuation in the Nirma Cement EIA

In a government-oriented system of valuation, monetary value is attached to most aspects of the natural environment. Through this, extraction becomes valued because it aids in generating revenue. Government-sponsored extraction of value then produces inequality to the extent that it is conducted through privatization and usually tied to unequal or undemocratic redistribution of benefits (Ayyub, 2014; Burke and Heynen, 2014). The EIA report in Nirma Cement specifically states that "the project will be contributing to the state exchequer, strengthening the self-sufficiency in cement in the country and if and when required, will earn foreign exchange in the future" (MIN MEC Consultancy Pvt. Ltd., 2008), revealing that profit is an important value. This indicates the EIA is a government-oriented system, where resources are managed for economic growth, tax revenue and public services directed at national priorities (Burke and Heynen, 2014; Kula and Evans, 2011). Thus, environmental issues raised in the EIA are minimized and considered mitigable (MIN MEC Consultancy Pvt. Ltd., 2008). The environment's value is viewed in economic terms, through the potential for future returns, which outweighs the project's impact on local communities and the environment.

Community members affected by the Nirma Cement project tended to value the environment differently, since they rely on it for their farming, firewood collection and main water source. EIAs tend not to examine this nuanced relationship, resulting in some impacts going unassessed. Community members are concerned about the negative impacts on agriculture, livestock and employment. For example, the expected increase in the salinity of the soil would be problematic because it would destroy farmers' ability to produce crops, possibly leading to permanent migration to find other work. As one woman in the community stated: "What about our children? Our animals? Our livelihood? We have been in this area for generations. We do not know how we will survive if we have to migrate." While the environment plays a role in livelihood production, there are more complex relationships with nature and resources that create a deep sense of place (Fernández-Giménez and Estaque, 2012). Valuing the land for providing a sense of place in combination with intimately knowing crop cycles, rain patterns and animal habits reveals a community's perspective as being part of a household-oriented value system (Pascua et al., 2017). This means that people's knowledge of the environment, accumulated through multigenerational experience and practice, is honoured (Burke and Heynen, 2014).

Since each valuation system focuses on particular aspects, it is important to recognize the subjectivity in valuation. Mostert (1996) points out that subjective valuation is as important as the potential environmental effects, particularly since projects are often introduced into regions where residents may already feel excluded from aspects of economic and social life (Walker and Reed, 2021). Decisions that are taken as a given often have an element of subjectivity and are not always based on scientific measures (Mostert, 1996). An example of this is the chosen study area, where companies tend to choose a smaller scoping area while community members usually want a larger area (Mostert, 1996). Community members impacted by Nirma Cement had indicated that a larger evaluation area would increase the number of villages that had to be examined, helping to illustrate further impacts to and connections between soil, animal and human health and social structures, beyond the immediate spatial boundary. Nirma Cement selected a relatively small area (10 km), resulting in a variety of villages not being assessed, despite the likelihood of them experiencing impacts.

Valuation in EIAs is also influenced by money, workers, commercial interests and the importance attached to public participation (Mostert, 1996). Defining nature and knowledge only in terms of market value erodes their character as public goods and reduces the possibilities for participatory engagement (Burke and Heynen, 2014). Thus, there needs to be multiple forms of knowledge that are valued so that local, experiential and scientific knowledges are engaged (Rice et al., 2015). An inclusion of more experiential ways of knowing enables and legitimatizes place-based meanings and provides culturally specific understandings (Rocheleau, Thomas-Slayter and Wangari, 1996). The government-oriented system is not inherently negative, but using only one system of valuation, particularly the dominant perspective, can lead to exclusion. Integrating traditional knowledge enables a holistic approach to adapting to changing environmental conditions (Fernández-Giménez and Estaque, 2012), and including diverse understandings of the environment creates a more comprehensive and legitimate EIA.

Value legitimization

Often, only certain knowledges are viewed as valid when valuing the environment and assessing impacts, as power discrepancies are present in knowledge and value systems (Rocheleau, Thomas-Slayter and Wangari, 1996). Currently, the predominant way EIAs measure the worth of the environment is based on scientific knowledge. Within this dominant narrative, local perspectives can be erased (Israel and Sachs, 2013) through a devaluation of knowledge, and this is reinforced through economic and legal processes (Burke and Heynen, 2014). If EIAs use dominant scientific narratives and women tend to be absent from these narratives (Harding, 1991), their values and understandings are automatically disregarded. This approach excludes uses of the environment that may not be documented and knowledge that has been passed down from generation to generation, as in the Nirma Cement case, where women's use of land, sense of place and concerns were not taken into consideration.

Many women in the affected communities viewed as fundamental their personal freedom and security when working on their farms and tending to their animals. Nirma Cement had brought in migrant workers to construct the plants, which had contributed to increased rates of violence against women in surrounding communities. Many women were worried that they would lose their freedom to walk alone between villages and farms or go outside alone at night.

Large-scale industrial development was also viewed as a threat to safe supplies of drinking water and traditional land use practices (working on the farm, grazing animals and collecting firewood), ultimately leading to an increase in male outmigration. However, women discussed that they were not against development in the area but wanted it to fit their needs, add value to their farming initiatives and not have adverse environmental impacts. Yet these views were not included in the EIA, which instead focused on the potential economic benefits of the affected communities.

Values shape processes of scientific reasoning (Druckman and McGrath, 2019). Thus, when scientific knowledge is legitimized by government, project proponents and EIA consultants through legal regulations, the values of these organizations become highly privileged and science's status is maintained as the largely uncontested authoritative basis for environmental decision-making (Burke and Heynen, 2014). Shifting to more inclusive EIAs would require questioning knowledge production systems (Rocheleau, Thomas-Slayter and Wangari, 1996). This would involve investigating and challenging deeply held social norms and power dynamics such as gender, caste and class, leading to robust and complex understandings of the environment (Hanson and Buechler, 2015). EIAs that consider and address these issues would legitimatize the values and enable increased participation of marginalized groups in formal environmental justice processes.

A promising tool for enabling this shift in values is GBA. The GBA framework recognizes the intersections of gender with diverse identity attributes, across policy, legislation and initiatives (Status of Women Canada, 2018). Broad guidelines include understanding GBA and the context in which a proposed project is situated, engaging early with diverse social groups, establishing a baseline and assessing potential effects across subgroups within communities (Walker and Reed, 2021). Using GBA can help build more inclusive EIA processes and reveal historical and contemporary injustices so that developments can be designed to avoid reinforcing pre-existing inequalities (Walker and Reed, 2021). For proponents this requires developing strong understandings of the local context and building opportunities for sharing best practices and experiences in facilitating inclusive engagement (Walker and Reed, 2021).

The type of knowledge that is valued plays a significant role in determining who participates and how participation is shaped in the EIA process. Currently in India, EIAs often focus on economic benefits, leaving out how communities want their resources used. As Walker and Reed (2021) identified, incorporating GBA can deepen analysis and point to more equitable mitigation and enhancement strategies.

From exclusion to collective movements

EIAs are a gendered process that produce gendered outcomes. Participation in EIAs is determined by rules, norms and perceptions (Agarwal, 2001) that define who should attend and speak up at meetings. In the public hearing for Nirma Cement, community members identified feeling that their voices and concerns were not being taken seriously or considered. Some women were also threatened, which discouraged their participation. When the valuation system is based on predominantly scientific and monetary values, this becomes a way to legitimize exclusion during the public hearing process. The creation of the Mahuva Andolan was a response to this exclusion and exclusion from spheres of power where environmental decisions are made (Rocheleau, Thomas–Slayter and Wangari, 1996). Thus, the (de)valuation of local knowledge within the Nirma Cement EIA forced women into informal spaces to ensure their voices would be included in the process.

The andolan played a role when formal spaces, even ones that have reserved certain places for women, such as panchayats (village councils), failed to incorporate and express their voices. As extensively documented by Agarwal (2000, 2001), panchayats have gender issues in that social and gender norms shape an individual's engagement. In Mahuva, although 33 per cent of panchayat positions are reserved for them, women often do not actively fill these roles. Their husbands often act as their informal surrogates, attending meetings, talking to community members and making decisions (Ghosh, Chakravarti and Mansi, 2015). Another norm that influenced women's exclusion are the times chosen for community meetings; they often occur at night when only male members can attend.

The andolan provided a space for women to feel more in control, allow them to represent themselves, present their environmental knowledge and communicate their messages (Mavroudi, 2008). An example of this was the 170-km foot march from the village of Doliya to Gandhinagar organized by the Mahuva Andolan in protest of the cement plant, in which over 5,000 people participated. As part of this protest, women made chapatis, onions and coconuts out of cement to give officials, specifically the Chief Minister of Gujarat, as a symbolic gesture to question the applicability of industrial development in the area. The women interviewed not only highlighted this activity but also stressed that they had been involved in every protest since the conception of the andolan.

The andolan proved to be an effective strategy – construction on Nirma Cement has not been completed and the case has been tangled up in the justice system for over 10 years – but participation has a cost, often paid by women. Many women indicated that to participate in andolan activities, they had to give up paid activity for the day and/or ignore important family responsibilities. This meant they were forced to decide between participating in environmental management and making ends meet. Thus, some women could only participate by providing support through cooking, watching children or tending to farms while other family members participated. This indicates there is still a significant gap between men and women in terms of experiencing the costs of participation (Agarwal, 2001).

Over time as the movement grew, strategic community alliances were built. A local politician who believed in not only the movement but also in protecting the environment was approached by the community to lead the andolan. This enabled the community, and subsequently women, to form a strategic alliance with and receive support from a political leader known for his work as a doctor in the villages, in bipartisan politics and as a supporter of both women and the environment. By inviting the politician, women redefined their leadership role to work alongside him – as mentioned above through the example of cement food – by creating their own initiatives within the larger collective's protests. This alliance enabled the andolan to extend its reach, allow community members to take their concerns to the Chief Minister of Gujarat and approach the court for intervention. However, the andolan is still a space of social relations and norms, influenced by the community it is based in (Agarwal, 2000). For example, because andolan meetings were often held at night – like other community meetings –women were unable to attend due to their care responsibilities. This highlights the gendered issues of participation in social movements and community decision-making, equity in the distribution of costs and benefits and the efficiency of protecting resources (Agarwal, 2000). Moving forward, for effective and gender-transformative change, women need more formal involvement, both in environmental social movements and in environmental management processes such as EIAs.

EIAs can legitimize diverse forms of knowledge and change predominant narratives by going beyond quantification of the environment solely for its monetary worth. This requires a commitment to incorporate pluralistic knowledge and encourage increased participation from communities whose ways of knowing and engaging with nature have traditionally been marginalized (Burke and Heynen, 2014). Without re-evaluating the role that systems of environmental valuation play in individuals' access to EIAs, women and other marginalized groups will consistently be excluded from participating. This exclusion then undermines procedural environmental justice, as marginalized groups will not be adequately represented in or able to fairly access environmental management spaces. Moreover, without protecting procedural justice, EIAs will most likely continue to fail to advance distributive justice, that is, the equitable distribution of environmental risks and benefits. The underlying value system of EIAs needs to encourage diverse understandings and challenge inequitable power dynamics. Otherwise, EIAs will remain a tool that favours whoever is validating the knowledge that is used.

A shift in values creates potential for change

Our values influence the choices we make and underpin societal institutions; what is valued in EIAs represents larger collective values about the environment. The Nirma Cement case provides a helpful analysis of how the embedded value systems within institutions shape how individuals participate in the management of their environment and resources. Dominant narratives about the environment, centred on cost-benefit analyses, can erase diversity of perspectives and knowledge, including those of women and other marginalized peoples (Israel and Sachs, 2013). This has led to a monopoly on who has the power to interpret and control the environment (Burke and Heynen, 2014). Thus, when there are contradicting value systems present, tensions regarding whose perspective and benefits matter most will arise.

The values underpinning EIAs influence the public hearing process, who can participate and what engagement looks like. When experts-only social and ecological knowledge guides political decisions about how landscapes will be transformed and for what ends, then knowledge hierarchies are likely to reinforce cultural hierarchies and material inequalities (Burke and Heynen, 2014). In the Nirma Cement case, the gendered nature of the EIA led to a shrinking of formal space and subsequent exclusion of women. Women's knowledge about the environment was not considered valid and their concerns were not taken seriously. Although the explicit and tacit processes of valuation led to a silencing of knowledge in the official EIA, informal spaces enabled the effective and powerful mobilization of the Mahuva Andolan, resulting in women's voices being heard.

However, many other cases do not result in positive outcomes for affected communities. Undertaking a GBA from the very beginning of the EIA process is vital to ensuring diverse groups are represented and any barriers to their meaningful participation are identified and addressed (Walker and Reed, 2021). This can help build inclusive EIA processes that create opportunities for enhancing distributive justice. Thus, a shift in values, which recognizes the importance of integrating a variety of knowledges, perspectives, understandings and approaches into EIAs through a combination of experiential and scientific knowledge, can help ensure that local experiences and diversity are authentically included. Such a shift could also increase the space available for all stakeholders and community members to participate in the broader political processes that affect their environments.

References

Agarwal, B. (1998) "Environmental management, equity and ecofeminism: Debating India's experience," *Journal of Peasant Studies*, 25, 55–95.

Agarwal, B. (2000) "Conceptualising environmental collective action: Why gender matters," *Cambridge Journal of Economics*, 24(3), 283–310.

Agarwal, B. (2001) "Participatory exclusions, community forestry, and gender: An analysis for South Asia and a conceptual framework," *World Development*, 29(10), 1623–1648.

Ayyub, B.M. (2014) "Systems resilience for multihazard environments: Definition, metrics, and valuation for decision making," *Risk Analysis*, 34(2), 340–355.

Bassett, T.J. and Peimer, A.W. (2015) "Dossier: 'À propos des relations natures/sociétés' – Political ecological perspectives on socioecological relations," *Nature Sciences Sociétés*, 23(2), 157–165.

Braun, V. and Clarke, V. (2006) "Using thematic analysis in psychology," *Qualitative Research in Psychology*, 3(2), 77–101.

Bulota, R. (2018) "Informal social networks as the precondition for mass mobilization: The case of Kaunas in the late Soviet period" in Piirimäe, K. and Mertelsmann, O. (eds.) *The Baltic States and the End of the Cold War*. Berlin: Peter Lang, 107–122.

Burke, B.J. and Heynen, N. (2014) "Transforming participatory science into socioecological praxis: Valuing marginalized environmental knowledges in the face of the neoliberalization of nature and science," *Environment and Society*, 5(1), 7–27.

Cornwall, A. (2002) *Making Spaces, Changing Places: Situating Participation in Development.* IDS Working Paper, 170. Brighton: Institute of Development Studies.

Diduck, A.P., Pratap, D., Sinclair, A.J. and Deane, S. (2013) "Perceptions of impacts, public participation, and learning in the planning, assessment and mitigation of two hydroelectric projects in Uttarakhand, India," *Land Use Policy*, 33, 170–182.

Dilay, A., Diduck, A.P. and Patel, K. (2019) "Environmental justice in India: A case study of environmental impact assessment, community engagement and public interest litigation," *Impact Assessment and Project Appraisal*, 38(1), 16–27.

Druckman, J.N. and McGrath, M.C. (2019) "The evidence for motivated reasoning in climate change preference formation," *Nature Climate Change*, 9(2), 111–119.

EPA (United States Environmental Protection Agency) (2014) "Methods, metrics, and indicators available for identifying and quantifying economic and social impacts associated with beneficial reuse decisions: A review of the literature," *Office of Research and Development*, Cincinnati, OH: EPA (United States Environmental Protection Agency).

Erlewein, A. (2013) "Disappearing rivers – The limits of environmental assessment for hydropower in India," *Environmental Impact Assessment Review*, 43, 135–143.

Fernández-Giménez, M.E. and Estaque, F.F. (2012) "Pyrenean pastoralists' ecological knowledge: Documentation and application to natural resource management and adaptation," *Human Ecology*, 40(2), 287–300.

Forsyth, T. and Walker, A. (2008) *Forest Guardians, Forest Destroyers: The Politics of Environmental Knowledge in Northern Thailand.* Seattle, WA: University of Washington Press.

Ghosh, R., Chakravarti, P. and Mansi, K. (2015) "Women's empowerment and education: Panchayats and women's self-help groups in India," *Policy Futures in Education*, 13(3), 294–314.

Government of Canada (2019) Impact Assessment Act: SC 2019, C 28, s 22. Available at: https://www.canlii.org/en/ca/laws/stat/sc-2019-c-28-s-1/latest/sc-2019-c-28-s-1.html (Accessed: 13 November 2020).

Gujarat Pollution Control Board. (2008) "Public hearing proceedings," Regional Office Gujarat Pollution Control Board.

Hanson, A.-M. and Buechler, S. (2015) "Introduction: Towards a feminist political ecology of women, global change, and vulnerable waterscapes" in Buechler, S. and Hanson, A.-M. (eds.) *A Political Ecology of Women, Water and Global Environmental Change.* New York, NY: Routledge, 1–16.

Harding, S. (1991) *Whose Science? Whose Knowledge? Thinking from Women's Lives.* New York, NY: Cornell University Press.

Hill, C., Madden, C. and Collins, N. (2017). *A Guide to Gender Impact Assessment for the Extractive Industries.* Melbourne: Oxfam Australia.

Howe, C.A., Butterworth, J., Smout, I.K., Duffy, A.M. and Vairavamoorhty, K. (2012). *Sustainable Water Management in the City of the Future: Findings from the SWITCH Project 2006–2011.* Delft, The Netherlands: UNESCO-IHE Institute for Water Education.

Impact Assessment Agency of Canada (2019) *Interim Guidance: Gender-Based Analysis Plus in Impact Assessment.* Available at: www.canada.ca/en/impact-assessment-agency/services/policy-guidance/practitioners-guide-impact-assessment-act/gender-based-analysis.html (Accessed: 13 November 2020).

Israel, A.L. and Sachs, C. (2013) "A climate for feminist intervention: Feminist science studies and climate change" in Alston, M. and Whittenbury, K. (eds.) *Research, Action and Policy: Addressing the Gendered Impacts of Climate Change.* New York, NY: Springer, 33–52.

Joshi, D. (2011) "Caste, gender and the rhetoric of reform in India's drinking water sector," *Economic and Political Weekly*, 46(18), 56–63.

Kula, E. and Evans, D. (2011) "Dual discounting in cost-benefit analysis for environmental impacts," *Environmental Impact Assessment Review*, 31(33), 180–186.

Mavroudi, E. (2008) "Palestinians in diaspora, empowerment and informal political space," *Political Geography*, 27(1), 57–73.

McFarlane, C. (2009) "Translocal assemblages: Space, power and social movements," *Geoforum*, 40(4), 561–567.

Mills, S., Dowsley, M. and Cameron, E. (2014) *Gender in Research on Northern Resource Development*. Gap Analysis Report #14. Whitehorse: Lakehead University and Yukon Research Centre, Yukon College.

MIN MEC Consultancy Pvt. Ltd. (2008) *Rapid Environmental Impact Assessment and Environmental Management Plan for the Proposed Cement Plant (Cement 1.91 MTPA, Clinker 1.5 MTPA), Captive Power Plant (50MW) and Coke Oven Plant (1.5 Lakh TPA) at Village PadhIarka, Taluka Mahuva, District Bhavnagar, Gujarat*. New Delhi: MIN MEC Consultancy Pvt. Ltd.

Morgan, R.K. (2012) "Environmental impact assessment: The state of the art," *Impact Assessment and Project Appraisal*, 30(1), 5–14.

Mostert, E. (1996) "Subjective environmental impact assessment: Causes, problems, solutions," *Impact Assessment*, 14(2), 191–213.

Paliwal, R.E. (2006) "EIA practice in India and its evaluation using SWOT analysis," *Environmental Impact Assessment Review*, 26(5), 492–510.

Pascua, P.A., McMillen, H., Ticktin, T., Vaughan, M. and Winter, K.B. (2017) "Beyond services: A process and framework to incorporate cultural, genealogical, place-based, and indigenous relationships in ecosystem service assessments," *Ecosystem Services*, 26, 465–475.

Perkins, P.E. (2007) "Feminist ecological economics and sustainability," *Journal of Bioeconomics*, 9(3), 227–244.

Rathi, A.K.A. (2017) "Evaluation of project-level environmental impact assessment and SWOT analysis of EIA process in India," *Environmental Impact Assessment Review*, 67, 31–39.

Rice, J., Burke, B. and Heynen, N. (2015) "Knowing climate change, embodying climate praxis: Experiential knowledge in Southern Appalachia," *Annals of the Association of American Geographers*, 105(2), 253–262.

Rocheleau, D., Thomas-Slayter, B. and Wangari, E. (1996) "Gender and environment" in Wangari, E., Rocheleau, D. and Thomas-Slayter, B. (eds.) *Feminist Political Ecology: Global Issues and Local Experience*. London: Routledge, 3–26.

Sainath, N.V. and Rajan, K.S. (2015) "Meta-analysis of EIA public hearings in the state of Gujarat, India: Its role versus the goal of environmental management," *Impact Assessment and Project Appraisal*, 33(2), 148–153.

Schlosberg, D. (2004) "Reconceiving environmental justice: Global movements and political theories," *Environmental Politics*, 13(3), 517–540.

Sheth, K.N. and Raval, K.C. (2014) "Relocation Nirma Cement Plant at Mahuva, Dist. Bhavnagar (Gujarat)," *Journal of Information, Knowledge and Research in Humanities and Social Sciences*, 3(1), 116–119.

Shrivastava, K. (2015) "Court cements Nirma claim" (blog), *Down to Earth*, 31 October 2010. Available at: https://www.downtoearth.org.in/news/court-cements-nirma-claim-2056 (Accessed: 12 August 2019).

Slootweg, F., Vanclay, F. and van Slootweg, M. (2003) "Integrating environment and social impact assessment" in H. Becker and F. Vanclay (eds.), *The International Handbook of Social Impact Assessment: Conceptual and Methodological Advances*. Northampton, MA: Edward Elgar Publishing, 56–73.

Srinivasan, B. and Mehta, L. (2003) "Using local knowledge" in H. Becker and F. Vanclay (eds.), *The International Handbook of Social Impact Assessment: Conceptual and Methodological Advances*. Northampton, MA: Edward Elgar Publishing, 161–178.

Status of Women Canada (2018) *Government of Canada's Approach: Gender-Based Analysis Plus*. Available at: https://cfc-swc.gc.ca/gba-acs/approach-approche-en.html (Accessed: 13 November 2020).

Walker, H. and Reed, M.G. (2021) "Assessing the intersections of sex, gender, and other identity factors in the new Canadian Impact Assessment Act" in Doelle, M. and Sinclair, A.J. (eds.), *The New Canadian Impact Assessment Act (IAA)*. Toronto: Irwin Press.

Wilkins, H. (2003) "The need for subjectivity in EIA: Discourse as a tool for sustainable development," *Environmental Impact Assessment Review*, 23(4), 401–414.

10 Holding international finance institutions accountable for environmental injustice

A case study of the Tata Mundra power plant in Gujarat

Avery Letkemann, Carinne Bétournay, Bharat Patel, Kirit Patel and Alan P. Diduck

Introduction

The state of Gujarat in western India has emerged as an industrial powerhouse in the last two decades. In 2017, the state contributed 18.4 per cent of India's total industrial outputs, the largest share among all of the states, despite being home to only 5 per cent of the nation's more than 1.3 billion people (Chakraborty and Basu, 2018; Pande and Datla, 2016). In order to address the increased energy requirements that came with this industrial growth, the Governments of India and Gujarat collaborated on the "24×7 Power for All" initiative (Government of Gujarat, 2016). This joint initiative commissioned the development of 4000–MW Ultra Mega Power Plants in nine locations across India (Buckley and Shah, 2019). Of these, two coal-fired plants were located side by side on the Kutch coastline near Mundra. Together, these two power plants burn more than 28 million tonnes of coal each year (Bhargava et al., 2012). Both plants have had significant impacts on the local communities and environment. The chapter examines the implications of one of the plants, the Tata Mundra power plant, for coastal communities in Kutch and the struggle of these communities in seeking environmental justice through local and international institutions.

Kutch, the largest district in Gujarat, is well known throughout India for the salt desert that covers 56 per cent of its land area (Rajgor and Rajgor, 2008). The coastal region of Kutch is an ecologically sensitive zone that boasts rich marine biodiversity. Large stretches of mangroves, coral reefs, mudflats, seaweeds, commercial fish and several rare marine species can be found in this area (Bhargava et al., 2012). Large-scale agriculture has not developed in the region because of arid conditions and high salinity of groundwater, but the local economies, which includes livelihood activities such as fishing, salt panning, small-scale farming and animal husbandry, have thrived here for generations (Bharwada and Mahajan, 2002; Gadh, Kara, Manjaliya and Reliya, 2011).

DOI: 10.4324/9781003141228-10

In 2001, the District of Kutch was struck by a 7.7-magnitude earthquake that claimed the lives of more than 17,000 people and destroyed much of the infrastructure that existed in the area (Lahiri, Jena, Rao and Sen, 2001; Simon, 2007). In the aftermath of the earthquake, the Gujarat government encouraged private sector investment in the region by creating tax incentives for the development of new industries [CGPL (Coastal Gujarat Power Limited), 2012]. Kutch is home to three of the nation's largest industrialized districts, and more than 70 large-scale and almost 30 medium-scale manufacturing industries have been established in Kutch since then (Pande and Datla, 2016; Patel, 2019). In 2004, the Adani Group, an Indian multinational conglomerate, established a Special Economic Zone and India's largest private commercial port on the Gulf of Kutch (Adani Group, 2019). Since this rapid industrial expansion in Kutch, significant efforts have been taken by local communities to protect livelihoods based on local resources and hold institutions accountable for environmental degradation. In efforts against the Tata Mundra power plant that extend over a decade, local communities continue to bear social, health and economic burdens caused by the plant without seeing any remedy in the foreseeable future for the restoration of the local environment.

Methodology

An in-depth qualitative case study method (e.g., Creswell, 2014) was adopted in this research. In order to understand the impacts on environmental justice in this case, the authors spent an extended amount of time in the field in India. The two lead authors from the Canadian team spent three months in coastal and inland villages around the Tata Mundra power plant. Another of the co-authors is based in Kutch and has been involved with the community in various capacities for the last decade, and provided expertise and organizational support to community-led struggles against industrial development in the region. An extensive review of the documents was conducted by the lead authors prior to arriving in India and throughout their time in the field. Ten semi-structured individual and two focus group interviews were conducted with key leaders, stakeholders and community members in the case, totalling 21 participants. The inclusion criteria included direct knowledge and experience of impacts caused by the Tata Mundra power plant or direct involvement in the struggle against the industry. Efforts were made to reach out to representatives of the Tata Mundra plant, international financing institutions and the local government to understand their perspectives on various issues, but responses were not received or interviews could not be arranged. The project-affected fishing communities of Tragadi Bunder and Kotadi Bunder were visited several times during the field research. We also consulted leaders of village panchayats (elected councils), non-governmental organizations (NGOs) and other development officials working in the region. Participant observations related to the local culture, Indigenous practices and gendered roles and the impacts of the power plant and other economic activities were recorded in field notes and were helpful for analysis and interpretation of qualitative field data. A grounded theory approach was taken in the preliminary analysis of the data (Birks, Chun Tie and Francis, 2019; Noble and Mitchell, 2016). Interview notes were read through a number of times to get a sense of the data set as a whole, and the results emerged from that analysis. Documentary and observational data supplemented this analysis and allowed for a meaningful understanding of the local culture, politics, livelihoods and resources of the area.

An overview of the Tata Mundra power plant

Coastal Gujarat Power Limited (CGPL), a wholly owned subsidiary of Tata Power Corporation Limited, won the bid in 2007 for establishing a coal-fired power plant in the region (TCE Consulting Engineers, 2007). The Tata Mundra power plant consists of five units that each produce 800 MW of electricity and in total occupy 1242 ha of common land provided by the government along the Kutch coastline near Mundra (Bhargava et al., 2012). This area includes 241 ha for the disposal of ash generated by the plant and 182 ha of land for constructing housing complexes for project employees and contract workers (TCE Consulting Engineers, 2007).

The total cost of the Tata Mundra plant is estimated to be US$4.2 billion and was financed through loans from the IFC (US$450 million), the Asian Development Bank (US$450 million), Korean credit export agencies (US$800 million) and local Indian banks (US$1.5 billion) (Bhargava et al., 2012). The electricity generated at the power plant is sold to several state-owned utilities that own power distribution grids in Gujarat, Maharashtra, Haryana, Rajasthan and Punjab (Bhargava et al., 2012).

The Tata Mundra power plant burns up to 12 million tonnes of imported coal each year. The coal is shipped to Mundra Port, owned by Adani Ports and Logistics, from Indonesia, Australia and South Africa. It is then transported on a 9-m-long (14.5 km) conveyor belt to the plant to be used in the operations (TCE Consulting Engineers, 2007). The fly ash generated by the plant is stored in silos and ponds on site [CGPL (Coastal Gujarat Power Limited), 2012]. The plant uses supercritical steam technology that is claimed to be an energy-efficient thermal system (Buckley and Shah, 2019).

An environmental clearance (EC) was awarded to the project by the Ministry of Environment, Forest and Climate Change (MoEFCC) in 2007 [MoEFCC (Ministry of Environment, Forest and Climate Change), 2007]. This initial EC included a yet-to-be-satisfied condition requiring the coal conveyor belt to be covered so as to reduce the dispersal of coal dust. The EC also required installation of a closed-cycle cooling system to reduce the use of water in the plant (TCE Consulting Engineers, 2007). Despite this requirement, Tata Corporation built the power plant using a once-through

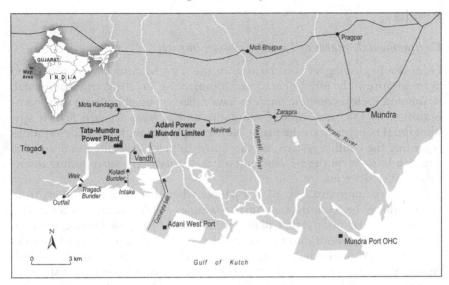

Figure 10.1 Map showing the location of the Tata Mundra power plant in relation to the Adani Power Mundra plant, the Adani West Port and several towns and villages in the vicinity. The coal conveyor belt and the intake and outfall channels are also represented. The arrows show direction of conveyor belt and flow of the water in the channels.

(open-cycle) cooling system. An intake channel 6.5 km long and 100 m wide was built on the coast to divert seawater to the cooling system. The water is released back to the sea by way of a separate 4.9-km-long outfall channel. The outgoing structure consists of a reinforced concrete channel, a pre-cooling channel, a weir and a dredged outfall that leads effluent from the thermal plant to the Gulf of Kutch (Figure 10.1). The plant uses 14.26 Mm^3 of water a day. As the warm effluent water disperses into the cooler seawater, it mixes to generate a thermal plume (TCE Consulting Engineers, 2007). Tata Power obtained a No Objection Certificate to use seawater from the Gujarat Maritime Board (GMB) in 2006 (CERC, 2018). The purpose of obtaining such a certificate is unclear and the legitimacy of the certificate itself is dubious, as the GMB is not an agency with an environmental mandate or expertise.

The Tata Mundra power plant is located directly adjacent to the Adani power plant, the largest single-location, private sector, coal-fired power plant in the world, and yet Tata Mundra was approved in 2013 without considering the cumulative impacts of the two plants. Although the Tata Mundra plant is the focus of this chapter, the cumulative impacts of both plants cannot be understated.

Results

Environmental, health and socio-economic impacts

From the beginning of construction of the Tata Mundra power plant, and more significantly after operations commenced in 2013, environmental, health and socio-economic burdens have fallen on neighbouring rural communities. At the same time, these communities have accrued few, if any, substantial benefits from the plant. Non-compliance with the EC conditions regarding the cooling system and the conveyor belt has resulted in noticeable negative impacts on the air, land and water quality of the surrounding townships and settlements.

There are approximately 10,000 people who earn their livelihoods from fishing along the Mundra and Anjar coastline. Fishing communities in this area practise a traditional type of foot fishing called *pagadiya*. During low tide, fishermen walk far out into the intertidal area to set up nets attached to sticks in the mud. At high tide, fish swim towards the algae-rich coastline to feed. As the water recedes, the fish get caught in the nets, allowing the fishermen to walk over and collect them (Gadh et al., 2011). Family members then process and hang the fish for drying along fences surrounding their homes.

One major concern for local residents has been the dredging of the intake and outfall channels for the power plant. Construction of the outflow has cut off access to the Tragadi and Kotadi fishing grounds, increasing walking distance for local residents by 4 km (7.6 km per fishing trip). The fishing communities believe that both channels are fundamentally altering the shallow marine ecology and threatening their livelihoods. Fishers suspect that large quantities of small, immature fish and eggs are pulled with the current into the channel and subsequently killed. Also, the water being discharged back into the ocean is warmer than the surrounding seawater by 5–7°C, extending several kilometres into the gulf [CAO (Compliance Advisor Ombudsman), 2013]. The outfall water contains a mixture of chemicals that have been added to prevent deposits of salt around the plant machinery. Over the last several years, fishers have noted visible depletions in the size, diversity, quality and quantity of the fish extracted from these waters. It has forced pagadiya fishers to go further into the open sea for their catch. This has made the purchase of small boats, diesel and additional equipment necessary.

Not only have fishing livelihoods been adversely affected, but also access to basic resources like water has been impacted. Dredging of the intake and outfall channels, along with dumping of dredging materials, has contributed to saltwater intrusion into the groundwater source that once satisfied local water needs (Gadh et al., 2011). The communities are now dependent on Tata for the provision of potable water. When that supply is not available, residents are forced to use their dwindling incomes to purchase water from often inconsistent supplies or otherwise continue to use the contaminated groundwater. The consumption of saline water is associated with numerous health complications, such as diarrhoea, abdominal pain, kidney failure and cardiovascular diseases, posing the greatest threat to children (Chakraborty et al., 2019). Since the water provided by Tata in some villages is not always sufficient to support the community, its distribution and usage have been a source of conflict in the past.

Another major cause for concern is the coal dust and fly ash that disperse through the air as a result of transporting and burning the coal. The company's contention is that the surrounding towns are not impacted by the ash due to the height of the 275-m tall chimneys. However, the height of the chimneys simply ensures that the ash is dispersed over a larger area, with neighbouring communities accumulating the largest quantities. Coal ash contains toxic heavy metals such as lead, selenium and mercury and is known to bioaccumulate in human and animal bodies (Hvistendahl, 2007). Residents from several villages have described experiences of waking up in the morning to find their homes and belongings covered in a thick layer of black ash. The ash from the power plant also contaminates fish laid out to dry in the fishing settlements. Concurrently, cattle rearers have also raised alarm bells regarding elevated rates of premature births and miscarriages among their livestock. This was rarely witnessed prior to the arrival of power plants in the region. Local villagers suspect that these unprecedented occurrences are, in large part, due to coal ash settling on grazing lands and subsequently ingested by the animals. People in surrounding villages have suffered a drastic rise in respiratory illnesses over the last seven years, with higher concentrations found among children and the elderly [CAO (Compliance Advisor Ombudsman), 2013]. Air pollutants also settle on salt panning sites along the coastline, a region which produces 30 per cent of the salt consumed in India (Bhargava et al., 2012). Considering that salt is produced in vast open plains susceptible to coal ash contamination and is subsequently exported throughout India, the public health implications are far broader in scope and involve a much wider population than simply those within the immediate area affected by the power plant's operations.

With coal dust and ash settling on crops and the rise of salinity in various irrigation sources, the quality, quantity and profits from agricultural activities have dramatically plummeted in the last ten years. As a result, many local people have sold their agricultural land or have been dispossessed of it by the state in order to create space for the Special Economic Zone and industries, including the Tata Mundra power plant. Farmers and pastoralists who persisted in trying to save their land-based livelihoods fear that they will not be able to make ends meet for much longer. The local villagers in Navinal were unanimous in stating that there are increasing levels of distress in the community and signalled an anticipated rise in suicide rates if the problems remain unaddressed.

The Tata Mundra plant has also disrupted the sociocultural fabric and traditional ways of life. Before the recent industrial development, men and women played complementary economic roles through their traditional livelihoods, and the opinions of women were listened to and respected. Since those traditional livelihood activities have been disrupted, opportunities for women to work outside of the household almost do not exist, especially for women who are older and uneducated. With these sources of income cut off, women are left with unpaid domestic labour, in what the CEO of a local women's organization described as a "step backwards for equal rights" for women.

Opportunities created by industrial development following the 2001 Kutch earthquake exist primarily for younger men. Community members commented that the opinions of elders (men or women) were no longer being heeded by younger people who are working for industry. Many male migrant workers have travelled to the area for work, often coming from far-off provinces in India without their families (Dutta, 2013). This has caused many women to express concerns for safety and social unrest. There is reluctance to encourage their daughters to attend school or go to work beyond their village or community. Thus, the impacts of the power plant are complex and far-reaching, affecting cultural identities and ways of life.

Seeking justice through local environmental regulatory institutions

Public hearings were held in 2007 as part of the environmental impact assessment of the project, but the hearings were sparsely attended because at that time few local residents were aware of the scope and magnitude of the potential impacts of the power plant. The struggle against the power project began in earnest in 2011, when construction of the intake and outfall channels began. Once they saw the size of the channels, residents began to realize how the environment and their livelihoods could be affected. It was then that local fishers started organizing their struggle under the auspices of Machimar Adhikar Sangharsh Sangathan (MASS), a local trade union that helps in selling fish in distant markets and advocates for the rights of fish workers in the District of Kutch.

Community leaders along with MASS met with representatives from the Tata Mundra plant twice in 2011. They requested that a bridge be built over the channels, since the traditional route taken to the *bunder* (a harbour used by fishers) and grazing lands had been cut off by construction of the channels. Their second request was for the company to comply with the original EC requiring the installation of a closed-cycle cooling system. Both of these requests were denied by the company, which claimed that a bridge and a closed-cycle cooling system were not feasible (Compliance Review Panel, 2013). The company's refusal to use a closed-cycle cooling system spurred representatives from MASS to write a letter to the Gujarat Pollution Control Board (GPCB). The GPCB is the agency charged with monitoring and enforcing environmental conditions stipulated in the EC and other pollution regulations in the state (Pande and Datla, 2016). Their letter was received but no action was taken by the GPCB to bring the power plant into compliance with the EC.

MASS pursued its concerns with the National Fishworkers Forum (NFF), a federation of small and traditional fish workers' unions in India. The NFF set up an independent fact-finding team to look into and report on the impacts of the power plant on fish workers' rights. The team was headed by a former chief justice of Sikkim and Rajasthan and a former chairperson of the Human Rights Commission of India. They visited the project-affected area twice, and the resulting report entitled *The Real Cost of Power* was released in June 2012. It provided recommendations for the Tata Mundra power plant, the Government of India, the Government of Gujarat and financial institutions involved in the project. The report was also submitted to the MoEFCC, which acknowledged the violation of the EC documented in the report and asked Tata Power to respond to each recommendation made by the report [MoEFCC (Ministry of Environment, Forest and Climate Change), 2012]. The community was never updated by the MoEFCC or the company on what actions were taken based on the report.

One of the reasons that Tata Power did not comply with the original cooling system condition was that it had quietly negotiated a revision of the condition from "closed-cycle cooling water system" to "suitable system" (CERC, 2018, 56). This amendment was later superseded when the federal government in 2015 passed stricter emission and technology standards [Environment (Protection) Amendment Rules 2015], requiring thermal power plants to reduce their water usage and use closed-loop cooling systems. Again, however, Tata Power negotiated an exemption, relying largely on economic grounds (i.e., increased production costs would lead to increased electricity prices) to convince the Central Electricity Regulatory Commission (CERC) in September 2018 to grant it an interim exemption from the new rules (CERC, 2018).

Seeking justice through the IFC's accountability mechanism

When it became apparent that Tata Power was not willing to address community concerns and that the actions taken in India were not successful in bringing the plant into compliance, MASS and other local leaders began to look for other means of seeking environment and social justice. It was at this time that local leaders were approached by the Centre for Financial Accountability (CFA), a New Delhi–based NGO, tracking the impacts of IFC loans in South Asia [CFA (Centre for Financial Accountability), 2021]. CFA members informed the community that IFC-funded projects are required to meet the IFC's internal Performance Standards on Social and Environmental Responsibility [IFC (International Finance Corporation), 2012]. The CFA offered help to the community and MASS leaders, if they were interested, in filing a complaint using the accountability mechanism of the IFC in Washington, DC. The community accepted the offer and accordingly submitted a complaint in June 2011 to the Compliance Advisor Ombudsman (CAO), an independent accountability mechanism set up for World Bank Group institutions, including the IFC. The complaint outlined the Tata Mundra power plant's numerous violations of the IFC's performance standards. The complaint also included violations of the EC issued by the MoEFCC.

In response to MASS's complaint, the CAO appointed a dispute resolution team. The CAO team visited the coastal settlement of Tragadi bunder twice in 2011 and spoke with the community and company representatives in the hopes of resolving the complaint without invoking a lengthy compliance investigation. Initially, the CAO's team offered to community representatives the options of sitting with Tata Mundra management to resolve the issue or pushing for a further investigation by the CAO [CAO (Compliance Advisor Ombudsman), 2012]. MASS organized coastwide consultation meetings with community members and shared the options available to them. Approximately 500 people attended these meetings, and it was eventually decided that they would not sit with the company and that they preferred that the CAO refer their complaint to its compliance division. Based on their field visits, the CAO team gave their initial assessment that the IFC had "failed to ensure the Tata Mundra project met the applicable environmental and social standards" (EarthRights International, 2017). The IFC rejected the CAO's initial assessment, which invoked a full compliance audit of the Tata Mundra project according to the IFC's institutional policies [CAO (Compliance Advisor Ombudsman), 2017].

The CAO completed the audit and disclosed its findings to the World Bank Group in October 2013. In response, the IFC released an action plan requesting the Tata Power Corporation address issues brought up in the complaint. To date, none of the actions outlined in the IFC's plan have been implemented. According to the IFC's policy, the responsibility for managing social and environmental risks and impacts is the responsibility of the client [IFC (International Finance Corporation), 2012]. The CAO's role is limited to preparing a report on whether a client is meeting the recommended action plan to meet the performance standards. If the client fails to comply, the IFC will work with the client to bring it into compliance to the extent feasible. If a client fails to establish compliance, the IFC will exercise remedies where appropriate. Accordingly, the CAO completed its first monitoring report in 2016, after which it carried out a monitoring visit in 2017 to evaluate Tata Power's compliance with the suggestions laid out in the action plan [CAO (Compliance Advisor Ombudsman), 2017]. Tata Power has shown little enthusiasm for the CAO's efforts over the years and has decided to pay the IFC loan in full (Patel, 2020). This action stripped the CAO of the limited influence it previously held. This was a strategic setback for community members, who had hoped that going through international institutions such as the CAO would be more effective than local and other domestic measures.

Seeking justice through litigation in the US courts

The work with the CFA and the CAO resulted in several opportunities for MASS to share its experience with other activists and organizations around the world. In 2014, leaders from MASS attended the World Bank Group Civil Society Forum to bring public attention to the struggle against the Tata Mundra power plant. The fishing community's story caught the attention of Washington, DC–based lawyers working with EarthRights International (ERI). After preliminary assessment of information provided by MASS, ERI offered pro bono legal assistance if the fishing communities and other stakeholders were interested in pursuing legal recourse in the United States. In April 2015, ERI lawyers launched the case of *Jam et al. v the International Finance Corporation* (Case 1:15-cv-00612-JDB) in the United States District Court for the District of Columbia. The plaintiffs included members of the fishing community, a local farmer, MASS and the panchayat of Navinal village. They sued the IFC over the destruction of livelihoods, loss and damage to their property and threats to their health caused by the construction and operation of the Tata Mundra power plant.

In response to the complaint, the IFC lawyers argued that under the International Organizations Immunities Act (IOIA), the IFC has complete immunity from legal actions in the US courts. This legislation was enacted in 1945 to protect international organizations to which the United States belongs from being sued in US courts. The act gives international organizations "the same immunity from suit and every form of judicial process as is enjoyed by foreign governments" (*Jam et al. v the International Finance Corporation*, 2015). When the act was first written, foreign governments did enjoy absolute immunity from suit in American courts, but that has since changed (Howe, 2019). In 2016, a judge ruled in favour of the defendants, concluding that the IFC was entitled to absolute immunity (EarthRights International, 2017).

The case worked its way up through the federal court of appeal, and in 2019, the Supreme Court delivered a landmark ruling in favour of the plaintiffs. The court concluded that international organizations engaged in commercial activity no longer held immunity from suit, as previously understood under the IOIA (Howe, 2019). The Supreme Court ordered the lower courts to adjudicate the case again, and in February 2020, the District Court upheld the IFC's immunity on the basis that the IFC's conduct was not carried out in the United States (Bold and Pereira, 2019; *Jam et al. v the International Finance Corporation*, 2015). This interpretation has almost made the Supreme Court's decision redundant with respect to the IFC, given that its mandate is to invest outside the United States. The ERI lawyers are considering appealing this decision; however, it will be a lengthy process and the community may exhaust its limited resources and its determination to fight.

Discussion: challenges and opportunities for seeking environmental justice

The concept of environmental justice emerged as a social movement in the 1960s and was originally conceived in the framework of racial inequalities (Perez et al., 2015). As the concept evolved, scholars underlined the need for a critical examination of environmental risks and benefits in society beyond racial inequalities of power (Miller, 1999; Rawls, 1971; Young, 1990). The distribution of risks and benefits of the Tata Mundra power plant, and industrial development as a whole in the region, calls for examining procedural, distributive, recognitional and restorative components of the environmental justice framework discussed in Chapter 1 of this book. Considering these components helps to understand the social, cultural and institutional context that results in distributive injustices, i.e., inequitable distribution of environmental burdens and benefits across gender, caste, class and livelihood portfolios (Schlosberg, 2004). Recognitional justice recognizes the diversity of participants affected by the Tata Mundra power plant and their varying capacity to respond to risks. Pre-existing sociocultural inequalities, based on caste and access to land and other resources, are the foundation for unequal capacity to cope with the environmental risks faced by the impacted communities in Kutch. Procedural justice focuses on participation in the institutional and political processes that implement environmental policy to ensure that there are opportunities for impacted communities to be involved in decisions that render environmental justice (Dilay, Diduck, and Patel, 2019; Schlosberg, 2004). Restorative justice has more recently been noted by scholars as a fourth integral aspect of environmental justice (Motupalli, 2018). It refers to the extent to which negative environmental impacts are remedied in a manner that takes into account the differing interests of the various stakeholders who have been affected.

Several steps taken by the impacted communities could have possibly resulted in an outcome that would have either closed the plant or forced it to comply with the EC conditions. The community leaders were clear that they did not want to stop the power plant from operating so long as it ensured restoration of natural resources on which their traditional livelihood activities were dependent. Our consultation with the impacted communities, MASS leaders and other activists, lawyers and development scholars revealed four suggestions for addressing the environmental injustice in this situation: restoration of the local environment; compensation for the loss of livelihood and other harms; creation of a community fund to address future health and well-being concerns; and changes to institutional policies to improve enforcement of EC conditions and financial accountability of international financing institutions. All of these suggestions reinforce the various elements of environmental justice, although implementation of the suggestions would require serious commitments to scientific and action research, community participation and institutional reform.

Environmental impact assessments are very important for identifying the adverse effects of a project and imposing EC conditions to ensure that those impacts are addressed. The impact assessment studies for the Tata Mundra plant were conducted inconsistently, and the EC conditions that were laid out originally were either ignored or amended several times under pressure from the project proponent [CERC, 2018; MoEFCC (Ministry of Environment,

Forest and Climate Change), 2007]. Instead of amending the EC, if the plant had installed a closed-cycle cooling system, the impacts of effluent water on the coastal ecosystem and livelihoods of pagadiya fishers would have been significantly lowered. Coverings enclosing the entire 14-km-long conveyor belt, coal ash ponds and coal transportation trucks could make a significant difference to air quality and other pollutants. Restoration of the environment to its prior state would require long-term interdisciplinary research into impacts and potential solutions for the revival of regional ecosystems. Kutch has very few research and environmental institutions that would have the capacity, human and financial resources and mandate to provide research support to the communities affected by rapid industrial development in the district. If the IFC is serious about enforcing its Performance Standards on Social and Environmental Sustainability, it should consider making substantial investments in independent research institutions in the region. In fact, the IFC anticipated serious negative social and environmental impacts of the project before making its lending decision and incorporated environmental guidelines into the loan agreement. However, it did not consider investing in local research institutions and the community to monitor and ensure compliance.

For ordering individual level compensation, the courts would require precise estimates of harm that can be directly attributed to the Tata Mundra power plant (Gibson and O'Faircheallaigh, 2010). However, the establishment of causal connections between power plant operations and harm to individual or community well-being is far beyond the capacity of local residents. The environment is degraded by a conglomerate of polluting industries in the region, not just the Tata Mundra plant. Furthermore, what fair compensation would look like, in terms of form and scale, has been a difficult question to answer for the impacted communities. The air and water polluted by the coal-fired power plants in the region can travel great distances, and people hundreds of kilometres away have been indirectly affected in ways that can be difficult to quantify. Recognizing the diversity of impacts and estimating the losses is a challenge that would require collaboration and an ongoing dialogue among various stakeholders, which the IFC, as one of the major lending institutions, should have instigated from the beginning of the Tata Mundra power project.

The creation of a community fund could be an effective solution for addressing environmental damage caused to natural resources such as fish stock, wildlife and biodiversity, grazing land and drinking water sources that are managed as common property resources (Gupta, 2003). The community fund could also help to recognize and address the risks that are difficult to attribute to a specific cause and effect phenomenon. The IFC and other lending institutions should make an initial corpus contribution to a community-managed fund. The sustainability of the fund could be ensured by requiring Tata Power to contribute a certain share of the profit earned every year from the Tata Mundra plant. The community should be empowered to manage the fund with support received by an advisory committee appointed by the IFC. The fund could finance services such as schools, hospitals, drinking water infrastructure, maintenance of water reservoirs and other commons for the community. The IFC showed some interest in establishing such community-level funding, especially after the US Supreme Court's decision that removed the absolute immunity that had been provided to the IFC. However, no concrete steps or initiatives have been taken in this direction.

As indicated in several other chapters of this book, an effective, fair and transparent impact assessment process could have prevented some of the procedural injustices that occurred at early stages in this case. For example, more and better public information about the power plant was obviously needed prior to the public hearings. Procedural justice was also compromised by failure of the MoEFCC, the GPCB and other environmental regulatory agencies to enforce compliance of EC conditions. The process for seeking justice through international institutions, such as the CAO of the World Bank Group or the US courts, offers promise but it needs to be more accessible and effective. Furthermore, seeking justice through the courts, either domestically or internationally, is challenging for local communities because most often they lack the scientific and legal resources to muster sufficient evidence to satisfy the legal burden of proof that they face. The accountability mechanisms available from financing institutions such as the IFC are important as alternative paths to seek justice. However, efforts of the community in this case have not met with any success. Though the CAO reports directly to the president of the World Bank Group, it only has the power to offer non-binding compliance reports. This lack of enforcement power inhibits the CAO from forcing borrowers to address environmental harm and social injustice. Institutional reforms in the World Bank's accountability mechanism are very important for environmental justice in India, as the IFC alone has invested close to US$15 billion in private sector–led projects in the country (World Bank, 2017).

Conclusion

The Tata Mundra power plant has had widespread adverse impacts on the surrounding environment, public health and socio-economic conditions. The project exists in an industrial hotbed and sits directly adjacent to the Adani coal-fired power plant. The cumulative impacts of industrial development in the area have not been assessed but have no doubt created injustices and erosion of community well-being stemming from the Tata Mundra plant, as outlined in the chapter. The Tata Mundra plant has been the subject of numerous complaints lodged with the accountability mechanism of the development bank along with a lawsuit in the American court system. These complaints centred around the plant's non-compliance with local laws and regulations, the social and environmental standards of the development banks and the EC issued by the MoEFCC. Suggestions have been made that could bring the plant into compliance and promote the four aspects of environmental justice (procedural, distributive, recognitional and restorative) in the region. Restoration of the natural environment would allow the surrounding communities to continue to make a living through the traditional means that have sustained them for generations. Adding a covering to the entirety of the 14-km conveyor belt that transports the coal from Mundra Port to the plant and converting the plant's open-cycle cooling system to a closed-cycle system would reduce the environmental pollution of the plant and impacts on public health. Providing compensation for loss of livelihoods and other harms caused by the plant and creating a community fund could be effective solutions for addressing environmental damage to natural resources and common property resources. Changing institutional policies for the issuance and enforcement of ECs, as well as the accountability of international financing institutions, could help to make justice-seeking processes more accessible to project-impacted communities. Pursuing environmental justice will be a long-term process that will require investment in research, planning, participation, the capacity to monitor progress and an ongoing dialogue between the affected parties. Considering the noted environmental, economic, social and health impacts on the communities surrounding the Tata Mundra power plant, meaningful dialogue is essential as a first step towards achieving lasting reparations for the people and rehabilitation of the affected ecosystems.

References

Adani Group (2019) *Mundra Port*. Adani Ports and Logistics. Available at: https://www.adaniports.com/Ports-and-Terminals/Mundra-Port.

Bhargava, S., Sampath, V., Bidwai, P., Ete, J. and Dutta, S. (2012) *The Real Cost of Power: Report from the Independent Fact-Finding Team on the Social, Environmental, and Economic Impacts of Tata Mundra Ultra Mega Power Project*. Available at: https://www.banktrack.org/download/the_real_cost_of_power/real_cost_of_power.pdf (Accessed: 26 September 2021).

Bharwada, C. and Mahajan, V. (2002) "Drinking water crisis in Kutch: A natural phenomenon?" *Economic and Political Weekly*, 37(48), 4859–4866.

Birks, M., Chun Tie, Y. and Francis, K. (2019) "Grounded theory research: A design framework for novice researchers," *Saga Open Medicine*, 2 January. Available at: doi:10.1177/2050312118822927.

Bold, P. and Pereira, B. (2019) *"Jam v. IFC – What Does It Mean for Accountability?" Independent Redress Mechanism*, 29 March. Available at: https://irm.greenclimate.fund/news/accountability-and-jam-v-ifc.

Buckley, T. and Shah, K. (2019) *Gujarat's Electricity Sector Transformation: A Role Model of India's Electricity Transition*. Institute for Energy Economics and Financial Analysis. Available at: https://ieefa.org/wp-content/uploads/2019/08/Gujarat-Electricity-Sector-Transformation_Final_August-2019.pdf.

CAO (Compliance Advisor Ombudsman) (2012) *CAO Assessment Report*. Available at: http://www.cao-ombudsman.org/cases/document-links/documents/TataDraftAssessmentReport_January_2012FINAL.pdf (Accessed: 26 January 2021).

CAO (Compliance Advisor Ombudsman) (2013) *Audit Report of IFC Investment in Coastal Gujarat Power Limited.* International Finance Corporation and Multilateral Investment Guarantee Agency, World Bank Group. Available at: http://www.cao-ombudsman. org/cases/document-links/documents/CAOAuditReportC-I-R6-Y12-F160.pdf.

CAO (Compliance Advisor Ombudsman) (2017) *India/Tata Ultra Mega-01/Mundra and Anjar.* Available at: http://www.cao-ombudsman.org/cases/case_detail.aspx?id=171.

CERC (2018) *CERC Appeal Judgement, Petition No. 77/MP/2016,* 17 September 2018. New Delhi.

CFA (Centre for Financial Accountability) (2021) *About Us.* Available at: http://www. cenfa.org/about/ (Accessed: 26 January 2021).

CGPL (Coastal Gujarat Power Limited) (2012) *Mundra Ultra Mega Power Project: Towards a Cleaner and Greener Future.* Available at: https://www.tatapower.com/plants-projects/ thermal-generation-projects/cgpl/pdf/umpp-greener-future121015.pdf.

Chakraborty, J. and Basu, P. (2018) "Linking industrial hazards and social inequalities: Environmental injustice in Gujarat, India," *International Journal of Environmental Research and Public Health,* 16(1). Available at: doi:10.3390/ijerph16010042.

Chakraborty, R., Khalid, K., Dibaba, D., Khan, M.D., Ahmed, A. and Islam, M. (2019) "Health implications of drinking water salinity in coastal areas of Bangladesh," *International Journal of Environmental Research and Public Health,* 16(19), 3746. Available at: doi:10.3390/ijerph16193746.

Compliance Review Panel (2013) *ADB Accountability Mechanism Complaint Form.* Available at: https://compliance.adb.org/dir0035p.nsf/attachments/complaint-mundra-ultra-mega-power-project.pdf/$FILE/complaint-mundra-ultra-mega-power-project. pdf.

Creswell, J.W. (2014) *Research Design: Qualitative, Quantitative, and Mixed Methods Approaches.* Los Angeles, CA: Sage.

Dilay, A., Diduck, A. and Patel, K. (2019) "Environmental justice in India: A case study of environmental impact assessment, community engagement and public interest litigation," *Impact Assessment and Project Appraisal,* 38(1), 16–27.

Dutta, S. (2013) *The Increasing Human Cost of Coal Power.* Supplementary Report to *The Real Cost of Power.* Available at: https://www.banktrack.org/download/the_increasing_ human_cost_of_coal_power/suplimentaryreportfinal.pdf (Accessed: 26 January 2021).

EarthRights International (2017) *Tata Mundra Coal Power Plant.* EarthRights International. Available at: https://earthrights.org/tata-mundra-coal-power-plant/ (Accessed: 26 January 2021).

Gadh, A., Kara, H., Manjaliya, J. and Reliya, H. (2011) *Tata Mundra CAO Complaint,* 11 June.

Gibson, G. and O'Faircheallaigh, C. (2010) *IBA Community Toolkit: Negotiation and Implementation of Impact and Benefit Agreements.* Toronto: Walter and Duncan Gordon Foundation.

Government of Gujarat (2016) *24X7 Power for All: A Joint Initiative of Government of India and Government of Gujarat.* Swarnim Gujarat, Gandhinagar. Available at: https:// powermin.nic.in/sites/default/files/uploads/joint_initiative_of_govt_of_india_and_ Gujarat.pdf.

Gupta, A.K. (2003) "Conserving biodiversity and rewarding associated knowledge and innovation systems: Honey bee perspective" in Cottier, T. and Mavroidis, P.C. (eds.), *Intellectual Property: Trade, Compensation and Sustainable Development The World Trade Forum,* vol. 3. Ann Arbor, MI: University of Michigan Press, 373–400.

Howe, A. (2019) "Opinion analysis: Justices hold that international organizations do not have near-complete immunity," *SCOTUSblog*, 27 February. Available at: https://www.scotusblog.com/2019/02/opinion-analysis-justices-hold-that-international-organizations-do-not-have-near-complete-immunity/.

Hvistendahl, M. (2007) "Coal ash is more radioactive than nuclear waste," *Scientific American*, 13 December. Available at: https://www.scientificamerican.com/article/coal-ash-is-more-radioactive-than-nuclear-waste/ (Accessed: 26 January 2021).

IFC (International Finance Corporation) (2012) *Performance Standards on Environmental and Social Responsibility*. International Finance Corporation, World Bank Group. Available at: https://www.ifc.org/wps/wcm/connect/24e6bfc3-5de3-444d-be9b-226188c95454/PS_English_2012_Full-Document.pdf?MOD=AJPERES&CVID=jkV-X6h.

Jam et al. v the International Finance Corporation (2015) *Class Action Complaint for Damages and Equitable Relief*. United States District Court, District of Columbia. Available at: https://earthrights.org/wp-content/uploads/ifc_tata_mundra_complaint.pdf.

Lahiri, A., Jena, P.R., Rao, R.K. and Sen, T.K. (2001) "Economic consequences of Gujarat earthquake," *Economic and Political Weekly*, 36(16), 1319–1332.

Miller, D. (1999) "Social justice and environmental goods" in Dobson, A. (ed.), *Fairness and Futurity: Essays on Environmental Sustainability and Social Justice*. Oxford: Oxford University Press, 151–172.

MoEFCC (Ministry of Environment, Forest and Climate Change) (2007) *Environmental Clearance Order, 4000 MW Mundra Ultra Mega Thermal Power Project by M/s Coastal Gujarat Power Ltd.*, 2 March. New Delhi: MOEF.

MoEFCC (Ministry of Environment, Forest and Climate Change) (2012) *Minutes of the 58th Meeting of Reconstituted Expert Appraisal Committee on Environmental Impact Assessment of Thermal Power and Coal Mine Projects*, 8 October. New Delhi.

Motupalli, C. (2018) "Intergenerational justice, environmental law, and restorative justice," *Washington Journal of Environmental Law and Policy*, 8(2), 333–361.

Noble, H. and Mitchell, G. (2016) "What is grounded theory?" *Evidence-Based Nursing*, 19(2), 34–35. doi:10.1136/eb-2016-102306.

Pande, R. and Datla, A. (2016) "Fighting pollution with data: Environmental audits and the Gujarat pollution control board." *Harvard Kennedy School Case Program, Case 2054.0*. Harvard University, Cambridge, MA.

Patel, P. (2019) *Agricultural Revision in Drought Prone Arid Region of Kutch: People Led, Market-Oriented Growth under Adverse Climatic Conditions*. 3rd World Irrigation Forum, Bali, Indonesia.

Patel, B. (2020) *Too Many Costs for Too Little: The Story of Tata Mundra*. Centre for Financial Accountability. Available at: https://www.cenfa.org/coal/too-many-costs-for-too-little-the-story-of-tata-mundra/.

Perez, A.C., Grafton, B., Mohai, P., Hardin, R., Hintzen, K. and Orvis, S. (2015) "Evolution of the environmental justice movement: Activism, formalization and differentiation," *Environmental Research Letters*, 10(10), 105002.

Rajgor, G. and Rajgor, M. (2008) "Women's perceptions of land ownership: A case study from Kutch district, Gujarat, India," *Gender and Development*, 16(1), 41–54. Available at: doi:10.1080/13552070701876144.

Rawls, J. (1971) *A Theory of Justice*. Oxford: Oxford University Press.

Schlosberg, D. (2004) "Reconceiving environmental justice: Global movements and political theories," *Environmental Politics*, 13(3), 517–540.

Simon, E. (2007) "State of play six years after Gujarat earthquake," *Economic and Political Weekly*, 42(11), 932–937.

TCE Consulting Engineers (2007) *Comprehensive Environmental Impact Assessment Study Report*. Coastal Gujarat Power Limited. Available at: https://www.tatapower.com/pdf/cgpl-mundra/CEIA-Complete.pdf.

World Bank (2017) *World Bank Group Finances*. Available at: https://finances.worldbank.org/Projects/IFC-Investment-Services-by-Country/ad4g-m4pt.

Young (1990) *Justice and Politics of Difference*. Princeton, NJ: Princeton University Press.

11 Environmental justice and participation for communities in southern India

Sand mining in Udupi District, Karnataka

Cassandra Szabo and Mahabaleshwar Hegde

Introduction

India has experienced significant economic and industrial growth in the last two decades and is now often referred to as an economic powerhouse. Despite this growth, and perhaps partly because of it, some of the country's most serious social, environmental and political challenges have worsened (Bhattacharya et al., 2015). These challenges include slow economic reforms, poverty, extreme inequality, poor education outcomes, poor healthcare, poor social indicators and inadequately prescribed policy solutions (Bhattacharya et al., 2015). Accompanying these problems is the concern that many of the industries contributing to the economy were, and still are, reliant on finite natural resources.

A large fraction of the poor in India rely on land- and natural resource-based livelihood practices, such as agriculture, fisheries, forestry, animal husbandry and related sectors (Venugopal, Appu and Gau, 2018; World Bank, 2001). Increased use of land and environmental resources by the state, industries and communities has, therefore, raised numerous conflicts in India. Studies suggest that India has more environmental conflicts than any other country in the world (D'Souza, 2012; Sarkar, 2018). These conflicts often pit economic gain against ecological and traditional uses and are frequently related to river water distribution, large dams, forest diversion, forest rights, displacements, industrial pollution and mining (Manupriya, 2016; Martinez-Alier, Demaria and Temper, 2014). India also has strict regulations under the Environment (Protection) Act 1986 to govern such environmental matters. However, the judicial system in India is overburdened by these conflicts and the public interest litigation that they attract (Whyte, 2011), and seeking justice can be a long process for litigants. India formed seven special administrative tribunals or judicial benches to resolve inter-state water disputes, along with the National Environment Appellate Authority, which is tasked with addressing environmental conflicts (Rosencranz, Sahu and Raghuvanshi,

DOI: 10.4324/9781003141228-11

2009). However, given the substantial number of unresolved conflicts, the National Green Tribunal (NGT) was set up in June 2010. It was enacted to allow for an efficient and expeditious disposal of matters related to environmental protection and conservation of forests and other natural resources (Gill, 2017; Kohli and Menon, 2012).

In the last decade, the Southern Zone Bench of the NGT has addressed more than 21 cases related to river sand mining from Karnataka state alone (NGT [National Green Tribunal], 2019). The growing construction and housing industries have created enormous demand for sand in Karnataka and elsewhere in the country. Studies have found that sand and gravel are the most extracted solid material globally and that the rate of extraction is far higher than its rate of renewal (Kohli, 2015; MoEFCC, 2016). Riverbeds, estuaries and coastal areas are the largest source of sand for industrial and construction activities. However, the excessive removal of sand, boulders and stones from rivers, creeks, estuaries and beaches can destroy riverbeds, affect fisheries and reduce clam collection in the affected region (Geological Survey of India, 2014; Jose, Shantanu and Venkatesh, 2014; Kohli, 2015). It may also cause river bank erosion, changes in river flow patterns and intrusion of saltwater into adjacent water bodies. In coastal areas, this can potentially lead to land erosion and intrusion of seawater during high tide.

Related to these environmental issues is a growing concern for environmental justice, the central tenet of which is ensuring that all individuals have equal ability to remain safe from environmental hazards and enjoy environmental benefits (Schlosberg, 2004). Environmental justice is a central feature of the United Nations Sustainable Development Goals (SDG), which are an important part of India's overall sustainability policy regime (United Nations, 2016). The current study aims to understand how India is approaching environmental justice, with a particular emphasis on the role of the NGT. The chapter presents a case study of sand mining in the Swarna River in Udupi taluka (block), Karnataka, and closely examines the NGT judgement of *Udaya Suvarna and Ors. v The Deputy Commissioner/Chairman District Sand Monitoring Committee and Ors.* (2016) to understand the NGT's role in promoting environmental justice. (Also see the summary of the case in Chapter 5, this volume.)

Background

The National Green Tribunal

The NGT was established by the central government in 2010 under the National Green Tribunal Act 2010 (Gill 2010, 2017). The tribunal replaced the earlier National Environment Appellate Authority and was given significant power and jurisdiction. The NGT was enacted to provide a specialized forum for effective and fast disposal of matters related to environmental protection, conservation of forests and damages caused to people or property because of violation of environmental laws or environmental clearance conditions (Shrotria, 2015). The main bench of the NGT was established in Delhi, with regional benches being created in the southern, central, eastern and western zones of the country. The NGT has the power to hear all civil cases relating to environmental issues and questions that are linked to the implementation of laws listed in Schedule 1 of the act, which include the Water (Prevention and Control of Pollution) Act 1974, 1977; the Forest (Conservation) Act 1980; the Air (Prevention and Control of Pollution) Act 1981; the Environment (Protection) Act 1986; the Public Liability Insurance Act 1991; and the Biological Diversity Act 2002 (Shrotria, 2015). The process for filing an application with the NGT includes completing an application form, which asks for basic information, a timeline of the issue and steps that have been taken by the applicant to try to remedy the situation. Claims for compensation can also be made, and such claims could involve various issues of damage to health, property or the environment (Shrotria, 2015). While the NGT is an important legal innovation in the country, the effectiveness and success of the tribunal is still in question (Gill, 2018), which points to the importance of examining it and its outcomes.

Environmental justice

Environmental justice provides a helpful framework for considering the effectiveness and success of the NGT. As noted earlier, the main principle of environmental justice is ensuring all individuals have equal ability to remain safe from environmental hazards and enjoy environmental benefits (Schlosberg, 2004). In addition to this distributive principle, several other aspects of environmental justice were important in guiding our analysis and discussion, including recognizing the differences among and within the communities affected (recognitional justice) and the degree to which people who are impacted are able to participate in the legal and political processes involved in environmental policy (procedural justice) (Schlosberg, 2004). Furthermore, as discussed in Chapter 1, this procedural aspect includes three important considerations respecting access to justice: capacity and standing to seek legal redress; access to legal proceedings that are fair, efficient and affordable; and access to mechanisms to enforce compliance with judicial decisions (Pring and Pring, 2009). Finally, our research considered restorative justice, because remediation of the environment and community, or lack thereof, can have direct implications for the equitable distribution of environmental costs and benefits (Rajan, 2014).

Sand and the environment

Sand is the world's most consumed natural resource after water and constitutes the largest volume of a solid mineral resource extracted and consumed globally (Peduzzi, 2014; UNEP, 2019). Studies have shown that the global consumption rate and demand for sand has increased rapidly and accounts for 47–59 billion tonnes annually (UNEP, 2019; USGS, 2013). Sand is a non-renewable resource that is used extensively in construction, the cement industry, glass manufacturing and preparation of concrete (Green Facts, 2018). The environmental and social impacts of sand mining are extensive and an issue of global significance (UNEP, 2019). The impacts of sand mining in river and coastal areas have direct, indirect and long-term implications (World Wildlife Fund, 2018). These major impacts include deepening of riverbeds due to erosion, loss of habitats, damage to biodiversity, alteration to hydrological functions of rivers and increased vulnerability to extreme weather events (Bayram and Önsoy, 2015; Campana, Marchese, Theule and Comiti, 2014). Studies suggest that large-scale extraction of sand at rates several times higher than its rate of natural replenishment has caused irreversible damage to land, biota and world's river systems (Peduzzi, 2014; World Wildlife Fund, 2018).

Sand mining in India

In India, sand mining is widespread and has become one of the largest threats to river basins. Increased demand related to rapid economic growth and inflated prices has created a black market for sand, leading to formation of a "sand mafia" (Kukreti, 2017). The mafia is usually associated with temporary contractors, suppliers and mining leaseholders who are mostly registered as unverified entities. These individuals and entities often have connections with local police, government and political officials (Romig, 2017). The Supreme Court of India reviewed the problems associated with sand mining in the case of *Deepak Kumar v State of Haryana* (2012) 4 SCC 629. The court acknowledged that sand mining was damaging the entire river ecosystem, including fish breeding areas, bird habitats and salinity levels in the rivers. The major impacts of sand mining in India are destruction of riverine vegetation, riverbed erosion, depletion of groundwater, increased salinity and turbidity and erosion of riverbanks (Geological Survey of India, 2014; Jose et al., 2014; Kohli, 2015). Besides the environmental damage, sand mining also affects the safety of bridges, structures and properties along the river by weakening of riverbeds (Kohli, 2015; Kukreti, 2017). Additionally, workers involved in sand mining have to deal with dangerous working conditions that involve diving in river water, living in unhygienic labour colonies and facing friction with local people (Romig, 2017). Villagers often actively try to stop sand mining because of its negative consequences and frequently use any avenue available to them, including protests, local governments and the NGT.

As a result of these impacts, sand mining in India is governed by several regulations and authorities. Though each state has a separate policy, at the federal level sand mining is regulated under the Mines and Minerals (Development and Regulation) Act 1957. As well, the Union Ministry of Environment

Forest and Climate Change (MoEFCC) issued Guidelines for Sustainable Sand Mining in 2016. Building on the guidelines, the ministry amended the regulatory provisions related to sand mining in the Environment Impact Assessment Notification 2006 (hereafter EIA Notification). In the amendments, the ministry attempted to decentralize regulatory powers by creating the District Level Expert Appraisal Committee (DEAC) and a District Environment Impact Assessment Authority (DEIAA), making these bodies responsible for assessing the impacts of mines (MoEFCC, 2016).

Methodology

The research followed a qualitative case study design (Creswell, 2014) aimed at understanding the overall process the villagers went through in engaging with the NGT, and the roles of other local, state and civil organizations. The primary methods were document review, semi-structured interviews and field visits. The document review focused on the NGT decision in the case of *Udaya Suvarna and Ors. v The Deputy Commissioner/Chairman District Sand Monitoring Committee and Ors.* (2016). The judgement and court proceedings were downloaded from the NGT website, and the environmental clearance was collected from the Karnataka State Environment Impact Assessment Authority. Details regarding the sand mining lease and approvals from district authorities were collected from the District Commissioner's office in Udupi.

The interviews and field visits were done from March 2018 to August 2018 using the participatory rapid appraisal (PRA) approach. PRA is particularly helpful for gathering information in a short period of time on local environmental and socio-economic conditions. It is a type of research that utilizes a systems perspective, meaning that the research and appraisal is done based on what participants in the system believe to be the critical elements, their relative importance and how the elements relate to one another (Beebe, 1995). The PRA approach was helpful in ensuring that the voices of those impacted by sand mining would be heard and the biases and voices of the researchers would be limited.

The selection of participants was done purposefully (Creswell, 2014). Given the sensitivity of the topic, only community members who were willing and able to engage were consulted. The most important information for the study was held by key informants in the village, and thus, those participants were contacted first. After that, a snowball approach was used in which the participants were asked to recommend other knowledgeable community members. Ultimately, 13 people were interviewed, including the petitioners in the NGT case and 6 other village members involved in the litigation. The average length of the interviews was 60 minutes. The interview data were recorded using a digital recorder and handwritten notes for those who were uncomfortable with the recorder. In addition, two focus groups were held involving five participants in the first focus group and four in the second. Those who participated included local leaders of the NGT case and impacted community members. In addition to interviews and focus groups, direct observations were made, and the data were recorded with photos and handwritten field observations.

Once the data collection was completed, analysis began in which themes and patterns (Creswell, 2014) were sought within the interview, focus group and observation data. The process involved transcribing the interviews and focus groups, reviewing the transcripts by coding data segments and grouping them, and identifying broad themes and patterns. MS Excel was then used to identify the number of times certain codes and themes appeared in transcripts, and these results supported the appraisal of the significance of themes. Overall, the analysis focused on evidence pertaining to the key concepts of environmental justice described above.

Case study: sand mining in the Swarna River

The Udupi District of Karnataka is located along the west coast and is separated from peninsular India by the Western Ghats mountain range. It lies between 13°04′ and 13°59′ latitude north and 74°35′ and 75°12′ longitude east, spanning an area of 3,575 sq. km. The Swarna River, one of the major rivers of Karnataka, flows through Udupi and holds special significance for the district. The river is a major drinking water source for Udupi city and an important irrigation source for rural Udupi District, providing water for thousands of hectares of farmland. With a total catchment area of 327 sq. km, the river originates in the foothills of the Western Ghats and flows for 62 km before reaching the Arabian Sea near the villages of Kotteshwara and Baikady (Kalra, Vasthare, Singhal and Udayashankar, 2019) (Figure 11.1). Baikady is situated on the riverbank about 13 km away from Udupi city. The geographical area of the village is 358.37 ha. In 2011, Baikady had a total population of 3,449, about 869 houses and an average family size of four (Indian Village Directory, 2018). Among the population of Baikady, 1,663 (48 per cent) are male and 1,786 (52 per cent) are female. Individuals from the general caste make up 92 per cent, while 7 per cent are from the Scheduled Caste and 1 per cent are Scheduled Tribes. A total of 60 per cent of the population depends on agriculture and grow coconut, areca nut, spices, vegetables and fruits. Around 20 per cent of the population depends on fishing and clam collection in the Swarna River and nearby coastal fishing grounds. The rest of the population are daily wage workers, small traders, auto or truck drivers and other professionals (Indian Village Directory, 2018).

Sandbar removal is an important activity in the Swarna River, and it can be done with low environmental impacts and be helpful for the local economy when it uses traditional methods and removes obstructions to navigation. However, intensive and large-scale sandbar removal, referred to as sand mining, has damaged the riverbed and fish breeding areas. In turn, this has seriously affected the income and availability of food for local villagers (Indian Village Directory, 2018). The natural filtration of saltwater in the estuary has also been affected, and as a result, village wells and ponds have become salty

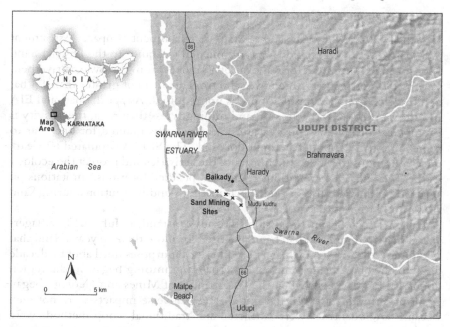

Figure 11.1 Map of the Udupi District, Karnataka, showing Baikady and key sand mining locations.

and unsafe to use for irrigation and domestic consumption. There has also been an increase in erosion on the banks of the river, due to an increase in the depth of the river and changes in flow patterns. A local villager provided a boat tour of the area, which showed the erosion taking place and where certain areas of land were sliding into the water (field notes, 12 July 2018).

Interviews with government officials shed light on how sandbars in the Swarna are identified and the procedures for applying for a sandbar removal permit. Officials explained that sandbars are identified via satellite images and GPS coordinates. They also explained that a seven-member committee meets and assesses whether an individual who has applied to remove a sandbar is eligible to do so (interview, Udupi, 8 July 2018). The officials said that no removal of sand takes place unless a permit is first issued, and they also reported that no sand mining was taking place in the region (interview, Udupi, 8 July 2018).

In the case that led to the petition to the NGT, a permit had been issued but there had not been a prior complaint from local people about navigation being obstructed. The permission to remove sandbars was, thus, partially based on a flawed rationale because sandbars can only be removed when they obstruct movement of local fishing vessels (as per MoEFCC Office Memorandum dated 8 November 2011). Furthermore, to comply with the requirement that local people be involved, a new association of dubious legitimacy was created to represent sand boat workers (interview, Udupi, 13 July 2018).

Seeking environmental justice

In April 2016, the Karnataka State Environment Impact Assessment Authority (SEIAA) granted an environmental clearance to the District Sand Monitoring Committee for removal of a sandbar from the Swarna riverbed. Permission was granted to remove sand from an area not greater than 1.4 ha, and therefore, public consultation was not required. As per the national EIA Notification 2006, if the area of sand mining is less than 5 ha, the activity is exempted from public consultation. Clearance was granted for one year to remove up to 14,700 metric tonnes of sand. The SEIAA stipulated 62 clearance conditions to reduce the impact of the activities and protect the ecology of the riverbed. The conditions related to specified locations, limitations on quantities removed, use of traditional methods and precautions during sand storage and transportation.

During a focus group discussion (Baikady tea stand, 14 July 2018), villagers stated that sand removal had been practised for more than 20 years. Sand that was removed had been primarily used for local purposes until about a decade ago, when some began to be sold. Actual sand mining began in the region about 10–12 years ago, when the Department of Mines and Geology began granting mining leases. In the initial stages, adverse impacts were not seen because low quantities were extracted and traditional scoop methods were used. In response to increased demand for sand for construction in larger cities, authorities started to give sand mining leases in larger numbers (interview, Udupi, 14 July 2018).

Villagers stated that sand mining started to occur in 30 designated areas of the Swarna River in 2014 and that many of these sites were mined without the required environmental and local approvals being in place. They also stated that because of poor monitoring by the regulatory agencies and limited accessibility of government records, the local villagers were not able to verify and distinguish between government-approved and illegal sand mining. Gradually, the impacts of intensive sand removal started appearing, and these sparked conflicts between local people and sand mining contractors. The major impacts that were noticed were erosion of riverbanks, increased turbidity of the water, change in river flow patterns and disruption of the river nutrient cycle. Fisherfolk claimed that changes in river dynamics affected fish breeding areas and subsequently affected the fisheries and clam harvests. Erosion of the riverbanks affected adjacent farmlands, and villagers noticed coconut trees falling into the river and water levels getting very deep (interview, Baikady village, 14 July 2020).

In response to these impacts, villagers mobilized farmers and fisherfolk and started approaching local authorities such as the village panchayat and the block-level Tahashildar (block administrative officer). Receiving little or no response at the local level, the community approached the District Commissioner of Udupi, who was also chair of the District Sand Monitoring Committee. Villagers believe that because of political demands and pressure from the sand mafia, no meaningful action was taken by the concerned authorities. In the absence of any tangible action, villagers organized public protests and sought the support of local media to publicize their concerns. However, the protests and media presence triggered violence, resulting in the protest leaders being arrested (interview, Udupi, 14 July 2020).

In a focus group discussion (Baikady village, 13 July 2018), participants stated that it was when the community ran out of administrative options that they decided to seek judicial redress through the NGT. During the discussion, it was evident that though community members were not aware of specific provisions, they were well aware of the existence of the Karnataka Sand Mining Policy 2011 and the Coastal Regulation Zone (CRZ) Notification 2011, which regulate sand mining in coastal areas. Initially, community members also had very little knowledge of the EIA process, environmental clearances and compliance with clearance conditions, but in interacting with their advocate throughout the litigation, they learned a considerable amount about these matters as they relate to sand mining.

The Udaya Suvarna case

The applicants who launched the NGT case were from Baikady village and were directly affected by the sand mining operations. These individuals had no prior experience with filing a lawsuit and even found it difficult to hire a lawyer. The applicants believed that the first lawyer they hired was bribed by the sand mafia; later they found a lawyer who was from the village who agreed to take on the case for a marginal fee. This lawyer came to the village to collect evidence from the villagers, who were actively involved by collecting documents related to the sand mining lease, panchayat approvals, the environmental clearance and the CRZ clearance (focus group, Baikady village, 13 July 2018). Most villagers did not go to the hearings, but they felt that the lawyer advocated for them well, seeing as he was from the area and shared their passion for keeping the village healthy (interview, Udupi, 14 July 2018).

The applicants started the litigation on 17 May 2016 in the NGT's Southern Zone Bench at Chennai. They argued that the Swarna River is the most important among the six rivers in Udupi District and that most of the downstream portion of the river falls under the CRZ, which is meant to protect coasts and promote coastal livelihoods (*Udaya Suvarna and Ors. v The Deputy Commissioner/Chairman District Sand Monitoring Committee and Ors.*, 2016). People who live on the banks of the Swarna River rely on the river for fisheries, clam collection, limestone collection and agricultural irrigation. Therefore, to ensure livelihood security for these communities, conserve and protect the coastal ecology and take into consideration sea level rise and global climate change, harmful activities such as sand mining in coastal and CRZ areas should be prohibited. The villagers informed the NGT that, as per the minutes of the District Sand Monitoring Committee, there were two sand processing units at Udupi/Kundapur District, namely M/s. Blue Sea Sand

Processing factory and M/s. Durga Parameshwari Industry. These units purchased sand extracted from the Swarna River and processed it for industrial sale. Therefore, the petitioners argued, the objective of sandbar removal that is permissible in the CRZ (i.e., promotion of coastal livelihoods) was totally defeated. Furthermore, the permit granted for removal of sandbars prohibited removal during the monsoon, as it overlaps with fish breeding season. Villagers also submitted photographic evidence in support of an argument that the sand miners violated the conditions of the environmental clearance granted by SEIAA. Additionally, based on the Karnataka SEIAA's meeting minutes, the applicants accused the SEIAA of not considering the sensitive river ecology and livelihood dependencies on the river.

Besides impugning the environmental clearance, the villagers also claimed that the Karnataka SEIAA and the District Sand Monitoring Committee did not monitor compliance and that this was in gross violation of the EIA Notification 2006. Owing to the lack of monitoring, an excessive amount of sand was removed and precautionary measures to protect the riverbeds were disregarded. This caused impacts on the natural habitat, fish breeding, clam fisheries and agricultural lands, all of which directly affected livelihoods of communities dependent on these resources.

The litigation cited several environmental laws and legal principles as well as articles from the Indian constitution. The applicants contended that the removal of sandbars was, in effect, illegal sand mining carried out in gross violation of the CRZ Notification 2011 and the Environment (Protection) Act 1986. They also argued that the mining was in violation of the conditions imposed by the MoEFCC in Office Memoranda dated 9 June 2011 and 8 November 2011. The applicants stated that the onus of proof was on the project proponents to show that their actions were environmentally benign, and if further deterioration of the environment was not stopped, long-term environmental degradation would occur. Additionally, they argued that applying the polluter pays principle meant that the project proponents should be held liable to make good on harms caused to the environment and private property. The applicants also contended that the application was filed to protect their rights as enshrined under Articles 21 and 48A of the Constitution of India and protect the broad public interest.

The Karnataka SEIAA responded by first admitting that as per the CRZ Notification 2011, mining of sand, rocks and other substrata materials was prohibited in the CRZ. However, they then asserted that the MoEFCC Office Memorandum dated 8 November 2011 stipulated the conditions for removal of sandbars, namely, that it must be done by traditional coastal communities and only by a manual method. The objective of granting permission and of the environmental clearance was to ensure that the only sandbars that could be removed were those that obstructed movement of fishing vessels of the local fishermen, as this would enable them to pursue their livelihoods in a safe manner. Furthermore, as per the EIA Notification 2006, sand mining within 5 ha was a permissible activity under Schedule 7 of the act. Therefore, the SEIAA argued, the environmental clearance issued for sand removal in the Swarna River was valid according to the prescribed environmental norms. In a rejoinder to the SEIAA, the appellants clarified that there was no sandbar deposition in the coastal rivers of Karnataka and that fishing boats are not common in those rivers. Therefore, they argued, it was clear that the provision for the removal of sandbars was being used as a ruse to mine sand, which is otherwise prohibited under CRZ Notification 2011.

After the final hearing, the NGT gave judgement in favour of the Baikady villagers on 27 February 2017. The court determined that the removal of sandbars was in fact being done illegally and gave various reasons, such as a faulty process in determining sandbar removal methods, exclusion of important stakeholders in the decision and failure to comply with the environmental clearance conditions. Thus, all sand removal from 2016 to 2017 using the existing clearance issued by the SEIAA was prohibited in the Swarna River and in the entire Udupi District. The NGT further stated that there cannot be any extraction of sand in the name of removing sandbars unless valid permits are granted in accordance with the guidelines as provided by the MoEFCC. It was decided that the State of Karnataka be in charge of the ultimate decision on issuing permits for sand removal, in strict compliance with the guidelines and conditions provided in the Office Memoranda of the MoEFCC (NGT [National Green Tribunal], 2019; Safi, 2017).

Although the NGT proceedings and the ultimate decision in the case were favourable to locally affected communities, advancing the litigation was fraught and practical application of the final ruling was problematic. One of the leaders in Baikady who was involved in the litigation was harassed by sand mining companies. He was also falsely charged, and eight arrest warrants were issued alleging that he created a public nuisance and disturbed the peace of the village. In some instances, the person had to hide and stay away from the village to avoid threats from the sand mafia (interview, Udupi, 14 July 2018). The villagers took immense risk in going against the sand mafia, and they described it as a brave act in pursuit of justice. Additionally, despite the NGT decision, villagers reported that the sand miners remained active and that there was little monitoring to ensure that the sand miners were complying with the judgement (interview, Udupi, 14 July 2018). Despite the fact that some sand mining was still occurring, the villagers mentioned that they had already witnessed fish returning to the waters, but that erosion was still occurring and there were still no clams to collect (interview, Udupi, 14 July 2018).

Discussion and implications: environmental and social outcomes

The NGT's final ruling in favour of community interests and environmental protection is consistent with a basic purpose of the NGT. Community members expressed positive feedback about the NGT process and the final judgement, although the harassment of a community leader marred the lead-up to the proceedings. Though only four members of Baikady village participated regularly in the NGT hearings and the entire process, other community members were aware of the court arguments and proceedings.

This case is important in terms of its environmental and socio-economic implications. First, the state of Karnataka has a stake in the removal and processing of sand because it promotes construction activities and economic growth. These are legitimate interests and in part explain why the sand removal permits were granted. Nevertheless, there still needs to be regulation of mining activities and compliance with environmental legislation and clearances in order to protect the broad public interest in a healthy environment. The NGT clearly recognized that in this case. Second, there were economic implications for the members of the village. Some villagers rely on the waters for fishing and livelihood activities, and some even work in traditional sandbar removal for their own local construction needs. Therefore, some villagers were strongly against the removal while others were supportive if it would help navigation and be conducted in the traditional manual fashion. Again, the NGT recognized these interests and protected them in its judgement.

Additionally, sand mining is quite sensitive in terms of regional politics, and there are political implications for how the state handles sand resources. For example, it was an important issue in the 2018 state election campaign, with some political parties saying they would encourage it and some saying they would advocate on behalf of communities being adversely impacted by sand mining. Since the state has a say and a stake in the extraction of sand, this further adds to the complications.

This case reiterates the importance of having avenues for public engagement in the court and political systems. Without the option of pursuing remedies through the NGT, the villagers would have had to keep hoping their protests and calls for help to the government would be answered. But with the NGT they were able to voice their concerns and have a say in what was happening to their community. This is crucial, seeing as the need for concrete and sand will likely only increase with India's rapid economic growth, meaning that these issues will continue to affect communities. It is, therefore, vital that the public be able to not only be engaged in these decisions but also have a powerful voice in them.

While the NGT did not focus on environmental remediation, and monitoring seems to have been lacking, what the court did allow was engagement of the villagers in decisions that were greatly impacting their lives. Nevertheless, even this level of procedural justice faced formidable barriers. Community members expressed the significant struggle they went through in finding a lawyer, in not having the funds to pay for a lawyer and then in being able to access court hearings and information. Additionally, with sand mining still occurring in the region, the community is struggling with how the court's decision will be enforced and how to restore the damage that was caused by the mining (interview, Udupi, 14 July 2018).

In addition to these procedural and restorative aspects, the other pillars of environmental justice were not well served in this case. Recognitional justice was not adequately addressed in that regulatory authorities did not fully acknowledge the impact of sand mining on local livelihoods. That is, local interests seemed to be subsumed in larger regional and national economic considerations. Similarly, distributive justice was not well served because without properly addressing impacts, the bulk of the negative consequences would have been felt at the local level, while the majority of the benefits would have been distributed to urban communities and the sand mafia. Additionally, the burdens of addressing the issue were unequal for the community as they needed to take considerable risks in order to address the mining. Restorative justice was also limited with little consideration being given to recouping the lost income of villagers, addressing environmental remediation and enforcing compliance with the NGT's order.

Although the NGT is not a foolproof solution for achieving environmental justice, in this case, it brought attention to the environmental inequities associated with sand mining, ruled in favour of community interests and environmental protection and revealed serious flaws in decision making by regulatory authorities. At the same time, this case shows there are substantial gaps in the administration and enforcement of NGT decisions. Left unresolved, these lacunae will undermine the advancement of environmental justice in India, as well as the country's pursuit of the United Nations Sustainable Development goal of providing access to justice for all.

References

Bayram, A. and Önsoy, H. (2015) "Sand and gravel mining impact on the surface water quality: A case study from the city of Tirebolu (Giresun Province, NE Turkey)," *Environmental Earth Science*, 73, 1997–2011.

Beebe, J. (1995) "Basic concepts and techniques of rapid appraisal," *Human Organization*, 54(1), 42–51.

Bhattacharya, S., Sharma, V.K., Dubey, R.K., Bhasin, T.M., Bhattacharya, D., Mulukutla, R.D., Mulukutla, S. and Aysola, U. (2015) "Some key issues and challenges facing India: Perspectives on policy and action," *Journal for Decision Makers*, 40(2), 209–241.

Campana, D., Marchese, E., Theule, J. and Comiti, F. (2014) "Channel degradation and restoration of an Alpine river and related morphological changes," *Geomorphology*, 221(September), 230–241.

Creswell, J.W. (2014) *Research Design: Qualitative, Quantitative, and Mixed Methods Approaches*. 4th edn. Los Angeles, CA: Sage.

Deepak Kumar v State of Haryana (2012) 4 SCC 629.

D'Souza, R. (2012) *Environment, Technology and Development: Critical and Subversive Essays*. New Delhi: Orient Black Swan.

Geological Survey of India (2014) *A Model Document on Impacts and Methodology of Systematic and Scientific Mining in the River Bed Material*. Available at: http://ismenvis.nic.in/Database/Scientific_Mining_of_the_River_Bed_Material_7425.aspx (Accessed: 16 November 2020).

Gill, G.N. (2010) "A green tribunal for India," *Journal of Environmental Law*, 22(3), 464–474.

Gill, G.N. (2017) *Environmental Justice in India: The National Green Tribunal*. Abingdon: Routledge.

Gill, G.N. (2018) "Mapping the power struggles of the National Green Tribunal of India: The rise and fall?" *Asian Journal of Law and Society*, 7(1), 1–42.

Green Facts (2018) *The Mining of Sand, a Non-Renewable Resource*. Available at: https://www.greenfacts.org/en/sand-extraction/l-2/index.htm (Accessed: 14 November 2020).

Indian Village Directory (2018) *Baikady*. Available at: https://villageinfo.in/karnataka/udupi/udupi/baikady.html (Accessed: 20 July 2020).

Jose, M.K., Shantanu, K.Y. and Venkatesh, B. (2014) "A study of effect of sand mining on riverine environment," *Hydraulics, Water Resources, Coastal and Environmental Engineering (HYDRO 2014)*, Bhopal, 18–20 December 2014 (special issue), *International Journal of Engineering Research*, (3), 2347–5013.

Kalra, K., Vasthare, R., Singhal, D. and Udayashankar, H. (2019) "Assessment of concentrations of trace elements in the groundwater sources along the course of Swarna river in Udupi District, Karnataka State, India," *International Journal of Civil Engineering and Technology*, 10(2), 1366–1375.

Kohli, K. (2015) *The Sand Mining Conundrum*. Available at: http://beta.indiatogether.org/sand-mining-conundrum-environment (Accessed: 16 November 2020).

Kohli, K. and Menon, M. (2012) "The nature of green justice," *Economic and Political Weekly*, 47(15), 19–22.

Kukreti, I. (2017) "How will India address illegal sand mining without any data?" (blog), *Down to Earth*, 30 September. Available at: https://www.downtoearth.org.in/news/flouted-with-impunity-58736 (Accessed: 16 November 2020).

Manupriya (2016) "India leads work environmental conflicts," *Scroll.in*, 15 April. Available at: https://scroll.in/article/806665/india-leads-world-in-environmental-conflicts (Accessed: 16 November 2020).

Martinez-Alier, J., Demaria, F. and Temper, L. (2014) "Social metabolism and environmental conflicts in India," *India@logs* 1, 51–83.

MoEFCC (Ministry of Environment, Forest and Climate Change) (2016) *Sustainable Sand Mining Management Guidelines*. Available at: http://environmentclearance.nic.in/writereaddata/SandMiningManagementGuidelines2016.pdf (Accessed: 16 November 2020).

NGT (National Green Tribunal) (2019) *National Green Tribunal*. Available at: http://www.greentribunal.gov.in (Accessed: 19 July 2020).

Peduzzi, P. (2014) "Sand, rarer than one thinks," *Environmental Development*, 11, 208–218.

Pring, G. and Pring, C. (2009) *Greening Justice: Creating and Improving Environmental Courts and Tribunals*. New Delhi: The Access Initiative.

Rajan, S.R. (2014) "Environmental justice in India," *Environmental Justice*, 7(5), 115–116.

Romig, R. (2017) "How to steal a river," *The New York Times*, March 2017. Available at: https://www.nytimes.com/2017/03/01/magazine/sand-mining-india-how-to-steal-ariver.html (Accessed: 20 July 2018).

Rosencranz, A., Sahu, G. and Raghuvanshi, V. (2009) "Whither the National Environment Appellate Authority?" *Economic and Political Weekly*, 44(35), 10–14.

Safi, M. (2017) "Villagers pay tragic price as Indian building boom drives demand for sand," *The Guardian*, 30 December. Available at: https://www.theguardian.com/world/2017/dec/30/india-sand-mining-conflict-deaths-building-boom-environmental-damage (Accessed: 1 August 2018).

Sarkar, S. (2018) "India has more environmental conflicts than any other country in the world," *The Hindu*, 2 June. Available at: https://www.thehindu.com/sci-tech/energy-and-environment/india-has-more-environmental-conflicts-than-any-other-country-in-the-world/article24066709.ece (Accessed: 16 November 2020).

Schlosberg, D. (2004) "Reconceiving environmental justice: Global movements and political theories," *Environmental Politics*, 13(3), 517–540.

Shrotria, S. (2015) "Environmental justice: Is the National Green Tribunal of India effective?" *Environmental Law Review*, 17(3), 169–188.

Udaya Suvarna and Ors. v The Deputy Commissioner/Chairman District Sand Monitoring Committee and Ors. (2016) Application No. 111 (SZ) and M.A. Nos. 133, 136, 138/2016 (National Green Tribunal).

UNEP (2019). *Sand and Sustainability: Finding New Solutions for Environmental Governance of Global Sand Resources.* Geneva, Switzerland: United Nations Environment Programme.

United Nations (2016) *Sustainable Development Goals.* Available at: https://sustainabledevelopment.un.org/?menu=1300 (Accessed: 10 August 2018).

USGS (2013) "Sand and gravel (construction) statistics" in Kelly, T.D., Matos, G.R. (eds.) *Historical Statistics for Mineral and Material Commodities in the United States.* Reston, VA: Geological Survey Data Series.

Venugopal, S., Appu, S. and Gau, R. (2018) "Adapting traditional livelihood practices in the face of environmental disruptions in subsistence communities," *Journal of Business Research*, 100(December), 400–409.

Why Sand Mining Is a Poll Issue in Mangaluru? (2018) (online video), added by Rajya Sabha TV, 30 April 2018. Available at: https://www.youtube.com/watch?v=RHj0z5tsI7s (Accessed: 16 November 2020).

Whyte, K.P. (2011) "The recognition dimensions of environmental justice in Indian country," *Environmental Justice*, 4(4), 199–205.

World Bank (2001) *World Development Report 2000–2001: Attacking Poverty.* Available at: https://openknowledge.worldbank.org/handle/10986/11856 (Accessed: 1 March 2018)

World Wildlife Fund (2018) *Living Planet Report 2018: Aiming Higher.* Grooten, M. and Almond, R.E.A. (eds.). Gland: WWF International.

12 Karwar fisherfolk's quest for environmental justice

Examining the roles of impact assessment, environmental regulatory agencies and legal institutions

Mahabaleshwar Hegde, Kirit Patel,
Alan P. Diduck and Debayan Gupta

Introduction: India's quest for port-led prosperity – the case of the Karwar port

The history of urban development in India and elsewhere illustrates the importance of economic advancement in regions close to seaports (Shan, Yu and Lee, 2014). Ports and coastal infrastructure projects have been considered gateways to rapid industrial production that is integrated with global supply chains (Aserkar, 2005; Munim and Schramm, 2018). As India strives to achieve the target of increasing its GDP to US$5 trillion by 2025, it has increased financial investments in port-based infrastructure projects linked to manufacturing and resource mining industries in the hinterlands. The Sagarmala is an ambitious port-led prosperity programme implemented by the Ministry of Shipping, Government of India. The programme, launched in 2015, has a 20-year plan that aims to modernize India's ports and enhance their capacity to import and export goods. The goal is to enhance port-linked industrialization and connect coastal infrastructure to hinterlands by including several new highways in the plan. Under the Sagarmala programme, four new ports and two expansions of existing ports have been proposed along the coast of Karnataka (Government of Karnataka, 2019). These projects are at different stages of planning and approval but have already created a stir among coastal communities in Karnataka. Fisherfolk along the coast have raised concerns, saying that the Sagarmala programme poses a serious threat to their fishing grounds (Deccan Herald, 2020). The fisherfolk from Karwar have been especially concerned, objecting to the proposed phase 2 expansion of the port in their town.

Karwar is a small town with a beautiful beach along the coast and unique views of the Western Ghats. It is a place where the Kali River, flowing from the Western Ghats, meets the sea and supports a productive and unique marine ecology. Scientists and the Karnataka State Coastal Zone

DOI: 10.4324/9781003141228-12

Management Authority (KSCZMA) declared the Kali River as a Critically Vulnerable Coastal Area (CVCA) in 2018 to ensure protection for its unique biodiversity, mangroves and marine environment (Down to Earth, 2018). Asia's largest naval base is located beside Karwar town. Close to 20 thousand families from the coastal area were displaced when the naval base was established near Karwar in 1994 (Equations, 2000).

The existing Karwar port was established in 1964 and is governed by the Karnataka Port and Inland Water Transportation Department. It has two berths and a capacity of 3 million tonnes per annum (MTPA). The government of Karnataka, under the Sagarmala project, proposed a Rs. 1900 crore plan to expand the Karwar port in 2017. The proposed project would add five berths to the port, which would enhance the existing capacity from 3 to 7.5 MTPA (Directorate of Ports and Inland Water Transportation, 2017).Despite opposition from local fisherfolk, the state government gave the project an environmental clearance (EC) in January 2019 [SEIAA (State Environment Impact Assessment Authority), 2019]. After a year of resisting the clearance, members of the fishing community petitioned the Karnataka High Court to protect the environment and their livelihoods. This chapter examines the struggle of the fisherfolk seeking environmental justice by questioning the project's environmental impact assessment (EIA) and the role of regulatory and scientific authorities. The chapter offers insights on various components of environmental justice compromised by a poor EIA and lack of opportunities for effective participation in regulatory and technical decision processes.

The environmental and social impacts of ports

Wang and Cullinane (2006) have reported that port activity enhances opportunities for manufacturing industries and, thus, becomes a core part of international trade. Several studies focused on maritime transportation, business and development sectors underline how ports play an integral role in the global supply chain and make direct contributions to the economic growth of a nation (Dwarakish and Salim, 2015; Sleeper, 2012; Vining and Boardman, 2008). The literature also highlights serious adverse impacts of such port-based economic growth on the coastal ecology, oceans and coastal communities (Damle, 2017; Heaver, 1993; Ng and Song, 2010). The construction of ports and related infrastructure has a massive adverse impact on coastal ecosystems and brings up significant socio-economic changes in the region (Naik and Kunte, 2016). Construction and dredging operations damage shorelines and marine habitat and biodiversity and impact the livelihoods of fisherfolk and coastal communities (Balaji, 2018; Boer et al., 2019; Damle, 2017;). The adverse environmental impacts of ports continue after completion of the construction phase. These include pollution due to handling of goods, oil spills during shipping and accidents and increased turbidity due to ship movements (Shirodkar, Pradhan and Vethamony, 2010). Several studies of the Karnataka coast (Hanumagond and Mitra, 2007; Hegde and Dinesh, 2016; Naik and Kunte, 2016; Sivadas, Gregory and Ingole, 2008) match the concerns identified above and call for action to protect costal environments and livelihoods. The Government of India has announced that environmental impacts of port construction, expansion and operation shall be assessed under the Environment Impact Assessment Notification 2006 (EIA Notification, 2006), Coastal Regulation Zone Notification 2011 [CRZ (Coastal Regulation Zone), 2011] and other regulatory processes. Despite such assessments, the impacts of the maritime sector are increasing, resulting not only in substantial damage to the coastal ecology but also in harm to livelihood activities along the coastline (Damle, 2017; Sivadas et al., 2008).

Methodology and analytical framework

This chapter draws upon the experience and learning from an action research project implemented by Namati India and the Centre for Policy Research (CPR), New Delhi. The lead author coordinated project activities involving fishing communities along the Karwar coastline. The Namati-CPR project promotes environmental justice by helping local communities that are affected by large infrastructure and industrial development to *know* the law, *use* the law and *shape* the law (CPR–Namati, 2017). Knowing the law implies helping the affected communities understand environmental legislation, processes, institutions and remedies. Using the law refers to supporting affected communities in identifying violations and procedural lapses and bringing them to the attention of regulatory and administrative authorities, along with credible evidence and documentation. Shaping the law entails enabling active participation of affected communities in decision-making and empowering them to demand positive legal and policy reform.

As part of knowing the law, we conducted in-depth analysis of policy documents using administrative, scientific and legal expertise. The documents were collected from various levels of the government, environmental protection enforcement agencies and scientific and legal institutions. We examined the project's pre-feasibility report and application form (Form 1), terms of reference issued by the State Environment Impact Assessment Authority (SEIAA), the EIA report, public hearing proceedings and minutes of the State Expert Appraisal Committee meetings. The EC documents issued by the SEIAA and the KSCZMA were studied in order to understand the environmental commitment made by the project proponent. We also examined various documents submitted by the government of Karnataka, to justify their decisions, as a defendant in the Karnataka High Court of case of *Bhaitkol Bandaru Nirashritara Yantrikruta Dhoni Meenugarara Sahakari Sangha v Karnataka Maritime Board*(2020). The extensive document review offered insights on impacts of port operations, baseline conditions and key indicators for monitoring valued environmental components and potential socio-economic, cultural and environmental impacts.

Around 8 focus group meetings and 14 interviews were conducted with community members, scientists, lawyers and government officials who were actively involved in public hearings, protests and seeking actions through regulatory and legal institutions. Several community-level meetings were also organized with the Uttara Kannada District Fishermen Association, Karwar to understand the collective concerns of the members about the project and their vision for coastal development and environmental protection.

The analysis presented in the last two sections of the chapter draws upon the environmental justice framework described in Chapter 1. Our analysis reveals how aspects of recognitional, procedural, distributive and restorative justice were impacted by the proposed port expansion project. To reiterate, recognitional justice emphasizes recognition of the diversity of participants, experiences and interests in communities that are affected by resource development and environmental governance decisions (Schlosberg, 2004; Williams and Mawdsley, 2006). Procedural justice calls for meaningful participation in the political and legal processes that create and manage environmental policy, including access to information and the courts (Gill, 2017; Pring and Pring, 2009). The consideration of equity in the distribution of the risks and benefits of resource development and environmental governance helps in assessing distributive justice (Schlosberg, 2004; Williams and Mawdsley, 2006). Finally, restorative justice focuses on the extent to which negative environmental and social impacts of resource development and environmental governance decisions are remedied (Motupalli, 2018; Rajan, 2014).

Results

EIA process for the expansion of the port: a cause of conflict

Consistent with the neoliberal policies evident in India over the last two decades (e.g., as described in Chapters 2 and 3, this volume), the Karwar port expansion project was visualized as a joint venture between the public and private sectors. It is common in such ventures for private sector partners to be secured after initial funding and ECs are obtained (Department of Economic Affairs, 2020). This was the approach seen in this case, with the Karnataka Directorate of Ports and Inland Water Transportation leading the development and submission of the Karwar port expansion proposal. As noted earlier, the project proposed construction of five berths, which would add to the two existing berths and create capacity to handle an additional 4.5 MTPA. The proposal included the reclamation of 17 ha of land from the sea by constructing an 800-m reclamation-bund with fill added behind it. Deepening of the navigation channel, for navigation of large vessels, would require dredging 19 m deep up to 12 km into the sea. The project would also require construction of a 1.35-km-long breakwater to reduce the intensity of waves hitting the berths.

In seeking the EC, the Directorate of Ports and Inland Water Transport prepared a 230-page EIA report with technical details of the project and its potential environmental impacts (Directorate of Ports and Inland Water Transport, 2017). The Karnataka Regional Pollution Control Board conducted a public hearing in February 2018. Despite being given little opportunity for obtaining information and having relatively limited capacity to respond to technical details, the local fisherfolk and citizens participated fully in the 6-hour hearing held at Karwar. They raised questions about land acquisition, the project's technical plans, including dredging of the 12-km navigation channel, and potential increases in water pollution and turbidity due to high traffic volumes. Their major concerns were cumulative impacts on marine life and shrinking of their fishing grounds and fish catches. Many of these fisherfolk had been relocated in 1994–1995 when the Government of India constructed a naval base south of Karwar port. In addition to the existing prohibition on fishing in a 15-km area around the naval base (Figure 12.1), fisherfolk have been subjected to massive land-use change over the last five years due to the expansion of National Highway 66 and construction of tourism facilities on the beach.

In addition to the concerns expressed by local fisherfolks, several studies (Naik and Kunte, 2016; Nayak, 2000, 2017) of the Karnataka coast have

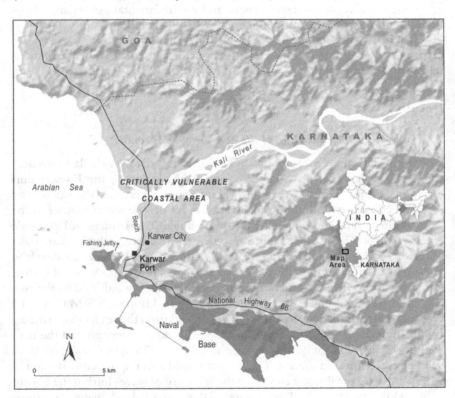

Figure 12.1 Map of the Karnataka coast, showing the Karwar Port.

shown that the port structures would have a major impact on the coast by altering beach erosion and accretion patterns. According to a recent research report submitted by the Regional Science Centre (Nayak, 2017), the breakwater of the proposed expansion project would increase the intensity of waves and flooding on the other side of the Karwar port, where Tagore beach and Karwar town are located. After heavy rain or high tides, Karwar's drainage system would be adversely impacted, and Tagore beach and other properties along the shore would face an increased risk of flash floods. Additionally, construction activity and dredging directly affect the seabed, fishing grounds and adjacent coastal hydrodynamics (Hegde and Dinesh, 2016). Naik and Kunte (2016) show that existing port structures and breakwaters in Karwar and the adjacent naval base are worsening coastal erosion. Beaches in Karwar have been subjected to intensive erosion since 2003 because of the construction of the naval base, dams and port (Hanumagond and Mitra, 2007). The rate of erosion has increased almost two times since 2000 and can be conclusively attributed to human activities such as construction of the naval base and operation of the port. Hanumagond and Mitra (2007) further ascertained that beach structures and sand dunes near the mouth of the Kali River were diminished from erosion caused by changes in wave and current patterns.

Despite the concerns expressed by the fisherfolk during the public hearing and the scientific reports noted above, the State Environmental Impact Assessment Authority fast-tracked the EIA process and issued an EC on 23 January 2019.

Challenging the validity of the environment clearance

As per the EIA Notification of 2006, how a project is categorized determines whether it will be appraised at the state or central level (see Chapter 6, this volume). Port projects with a capacity of less than 5 MTPA fall into Category B and are, thus, appraised at the state level [EIA Notification 2006, Schedule, entry 7 (e)]. In this case, the government of Karnataka strategically chose to limit the size of the port expansion to less than 5 MTPA, so that the project proposal would be subjected to the Government of Karnataka EC process, managed by the SEIAA, rather than the federal process administered by the Ministry of Environment, Forest and Climate Change (MoEFCC) in Delhi. In addition to project categorization, the general conditions specified in the EIA Notification of 2006 have relevance for the Karwar port expansion because the project could potentially impact protected areas. As per the general condition, any coastal development project within a 10-km radius of an ecologically sensitive area declared for marine conservation must be appraised at the federal level even if it would otherwise have been a Category B project. However, in this case, the project proposal was appraised by the SEIAA, and so the fishing community, led by the Uttara Kannada District Fishermen Association, approached the MoEFCC in May 2019 and requested that it revoke the state-issued EC.

The fisherfolk argued that the proposed project is located within 3 km of the Kali River, which under the approved Coastal Zonal Management Plan is an ecologically sensitive or protected area [KSCZMA (Karnataka State Coastal Zone Management Authority), 2018]. Therefore, according to the general clause of the EIA 2006 referred to earlier, the SEIAA lacked jurisdiction to appraise the environmental impacts of the Karwar port expansion project, and the EC issued by the SEIAA should be revoked. This argument is in line with the recent judgement issued by the National Green Tribunal in the case of G. *Sundarrajan v Union of India and Others*(2018), where the EC issued by the state government's agency was revoked.

Another ground for challenging the EC relates to a lack of attention to cumulative impacts, a common criticism made by community activists, scientists and EIA scholars about EC proceedings (Agrawal, Lodhi and Panwar, 2010; Grumbine and Pandit, 2013; Menon and Kohli, 2007; Rathi, 2017). In this case, the Directorate of Ports and Inland Water Transport argued that the proposed project has only a 4.5-MTPA capacity, and this was accepted by the SEIAA, which seems to have failed to consider the obvious fact that the project would add to the port's existing capacity of 3.0 MTPA, creating a total capacity of 7.5 MTPA.

Additionally, the Uttara Kannada District Fishermen Association argued that the project proponent provided false and misleading information to obtain the EC from the Karnataka SEIAA. According to section 8 of the EIA Notification, which prohibits deliberate concealment of information or provision of misleading data, this should result in the cancellation of the EC. The fisherfolk argued that, in addition to concealing information about the ecologically sensitive zone in the vicinity of the project, the EIA report failed to provide information on the existence of the unique ecosystems and rich diversity of aquatic species on the Karwar coast. For example, the Central Marine Fisheries Research Institute has documented the presence of Indo-Pacific dolphins in inshore waters of Karwar (Vaidya, 2011; Vivekanandan and Jeyabaskaran, 2012). Similarly, the Kali River estuary, located close to the project site, is recognized as an important breeding and nursery grounds for juvenile fish species (Haragi and Naik, 2012; Naik, Rathod and Bhat, 2005; Shaikh, Rathod and Durgekar, 2017). Moreover, other studies have reported the assemblage of economically important as well as ecologically significant marine species on the Karwar coast (Bandekar, Neelkantan and Kakati, 2011; Kaladharan, Zacharia and Vijayakumaran, 2011; Tenjing et al., 2018).

The EIA report also failed to consider the risk factors ascribed to port activity such as coastal erosion, turbidity due to dredging, pollution by shipping activities and oil spills. This is a major flaw given that, as noted earlier, existing industrial activities, construction of the naval base, port activities and ship accidents have for years been causing damage to the marine environment of the Karwar coast (Hanumagond and Mitra, 2007; Sivadas, Gregory and Ingole, 2008). For example, a large oil spill occurred near the Karwar coast in 2006 as a result of an accident involving the bulk carrier MV Ocean Seraya

(Sivadas et al., 2008). Making matters worse, ship accidents and oil spills near the port generally happen with the strong winds in the monsoon season, and this overlaps with the breeding season of marine fishes (Sivadas et al., 2008).

The final EIA report, prepared by the Karnataka Directorate of Ports and Inland Water Transportation after public hearings and other consultation, does not refer to any of these studies and hitherto has remained silent on the state of marine biodiversity and ecological resources. Because of the foregoing gaps in the EIA report, the potential cumulative impacts of the port expansion were not sufficiently addressed during the EIA review process.

Conflicting roles of the coastal zone authority

The federal office of the MoEFCC acknowledged the concerns expressed by the fishermen association and sent a letter to the Karnataka SEIAA in June 2019 (Huralimath, 2019). It asked the SEIAA to "reconsider the decision and take appropriate action with respect to the complaint" received. The lack of any response from the SEIAA implied that the Karnataka government either revoked the EC or put a hold on it until the concerns expressed by the Karwar fisherfolk were satisfactorily addressed. While the fisherfolk were waiting for a resolution, the KSCZMA granted a separate clearance to build a 1.3-km-long breakwater on 30 December 2019 [KSCZMA (Karnataka State Coastal Zone Management Authority), 2019]. The port authority immediately started construction in December 2019 without obtaining from the Karnataka State Pollution Control Board (KSPCB) a Consent to Establish under the Water Act 1974 and the Air Act 1981. These consents are required for large infrastructure projects such as the port expansion. The port authority claimed that the breakwater was a separate activity from the port expansion for which the EC was still under review. The KSCZMA endorsed that position and seemingly presumed that their approval of the breakwater would have no repercussions for the pending decision on the EC decision or the marine environment. This decision by the regulatory agencies disappointed the local communities and resulted in extensive protests along the coast.

The KSCZMA is the primary environmental regulatory agency constituted under the Coastal Regulation Zone Notification 2011, with a mandate to protect and improve the quality of the coastal environment. It is not a managerial branch run by administrative bureaucrats but a body with scientific expertise from various disciplines. It is aberrant that the KSCZMA considered construction of the breakwater as being separate from the port expansion. The clearance letter did not contain any conditions or information about the existence of the legally notified CVCA. This omission is surprising given that the KSCZMA played a pivotal role in the decision-making related to identification and declaration of the CVCA near Karwar port [KSCZMA (Karnataka State Coastal Zone Management Authority), 2018]. Furthermore, issuance of an independent clearance for breakwater construction while the EC for the port was being reviewed by the SEIAA questions the KSCZMA's interest in or ability to conduct a comprehensive assessment of the port project.

Likewise, the SEIAA ignored the concerns raised by the fishing community and scientists on the potential cumulative impacts of the port on the CVCA located just 3 km away. Fisherfolk emphasized that besides disregarding the ecological importance of the region, the project proponent also concealed information about pending land dispute lawsuits affecting the project site. In the Form 1 (application) for the project, the proponent claimed that there was no pending litigation concerning the land the project would use. Fisherfolk pointed out that in fact there were three pending lawsuits (*Krishna Vithoba Javekar v State of Karnataka and Others*, 2018). The failure of the SEIAA to acknowledge these disputes in the EIA report calls into question the agency's credibility and transparency.

Finally, the KSCZMA and other such authorities involved at the state level cannot attribute lapses in performing their duties to a lack of interagency communication. The KSCZMA, the SEIAA and other environmental agencies involved in granting clearances and consents come under the purview of the Karnataka Department of Forest, Ecology and Environment. All these agencies have several overlapping members and are often overseen by the same senior-level officer from the department office in Bangalore.

Implications for environmental justice

The fisherfolk and other communities in Karwar lost their patience once the Directorate of Ports and Inland Water Transportation started construction activities in January 2020. The Uttara Kannada District Fishermen Association launched aggressive protests involving fishing communities along the Karnataka coast and called for a bandh(general strike). The bandh, which lasted for 15 days, involved large rallies and drew a violent response from police that included detention of fisherfolk. The fisherfolk approached the Karnataka High Court, with the help of a lawyer from Bangalore, to stop construction of the breakwater. The Karnataka High Court issued a stay order on the construction activities on 23 January 2020. The fisherfolk hold the state environmental agencies responsible for the eruption of the protest, damage caused to public property, loss of livelihood income and the subsequent litigation in various courts. The High Court hearing began in October 2020, with the SEIAA and other government agencies arguing that the EC issued for the project was valid, has not compromised any procedural protocol and is in accordance with the EIA Notification 2006. They also argued that construction of the breakwater was a separate activity and not subject to the EC. Instead, they argued, they only need the Consent to Establish from the KSPCB and a Coastal Regulation Zone clearance from the KSCZMA. The court has initially questioned these claims and has extended the stay order for the construction activities. It also asked the government to submit further information about its various claims and on the decision-making processes that were followed. Whether the litigation will lead to a permanent and equitable solution for the fishing community, only time will tell.

Irrespective of the outcome from the litigation, procedural injustices in this case arose early and often. One of the first problematic issues was the question of whether Karnataka State had the authority to conduct the EIA appraisal. This fundamental jurisdictional issue created conflict and cast doubt on the validity of the entire EIA process. However, even if the project proposal had been appraised by the central government, it is not certain the outcome would have been different for the fisherfolk and local communities. ECs issued by the central government also face criticism for a lack of community engagement in decision-making (see Dilay, Diduck and Patel, 2020; Hegde, Patel and Diduck, 2021).

Procedural justice starts with the submission of truthful information and collection of quality data at the beginning of the EIA process for any project (Banham and Brew, 1996). The lack of integration of the port's various components, such as the existing port and the breakwater construction, prevented a comprehensive study of collective impacts. Activities such as dredging, land reclamation and seabed filling are known to cause massive impacts on fishing grounds and the marine environment. However, the fact that little attention was given to these impacts suggests there was little recognition of the varied communities and interests affected by the project and of how impacts would be distributed. This suggestion is supported by the fact that impacts on the notified ecologically sensitive area, the Kali River estuary and the habitat of marine mammals in the Karwar region were not ascertained during the entire EIA study. Persistent disregard for the eco-sensitive areas in the region throughout the EIA process undermines the objective appraisal of the project. Thus, lack of information in the EIA report on important marine organisms, the ecological significance of the coast and the cumulative impact of project activities raises a concern about the overall quality of the EIA. Having public consultations and an expert appraisal based on a flawed EIA report demonstrates the depth of the procedural dodges that occurred during the EIA process.

The cost-benefit analysis reported in the EIA report failed to recognize the loss of livelihoods due to damage to fishing grounds and tourism beaches. Additionally, a large section of a fishing village that would be needed to be evacuated was not recognized in the EIA study. These flaws are consistent with reports in studies that project costs and timelines are systematically underestimated and benefits are overestimated (e.g., Damle, 2017). The Karnataka SEIAA's decision to approve the project despite the objections raised in the public hearing by the local communities casts doubt on the legitimacy of the hearing and the integrity of the EIA process. Procedural anomalies that occurred during the EIA were apparent from the EIA report, public hearing meeting minutes and State Expert Appraisal Committee proceedings. Such flaws create barriers to achieving procedural justice, which in turn hold back the recognitional, restorative and distributive aspects of environmental justice. Without fair and open procedure, it is virtually impossible to have due regard for the diversity of actors and interests involved, achieve an equitable distribution of the costs and benefits of a development decision or craft an equitable and effective remedial plan.

The failure to consider recognitional and distributive justice prevented the advancement of restorative justice. The KSCZMA's granting of a separate Coastal Regulation Zone clearance for the construction of the breakwater without imposing any meaningful environmental conditions indicates a lack of thoughtful environmental decision-making. Even the EC issued by the state environment agency did not stipulate conditions to protect the ecologically sensitive zones. Restorative measures to address the ramifications of port activities, such as pollution, coastal erosion and damage to coastal drainage systems were not sufficiently addressed in the EC letter.

Who will stand with rural communities?

It is apparent that neither the EIA report nor the EC conditions addressed the concerns of local communities about protecting their environment and livelihood. The top-level decision-making committees of the MoEFCC at the state as well as federal levels in the neoliberal era tend to disregard concerns of project-affected communities, viewing them as "politically motivated." In these circumstances, the responsibilities of scientific and environmental regulatory agencies, such as the KSCZMA and KSPCB, are intensified when it comes to ensuring objective and fair EIA and EC processes. These institutions are equipped with specific scientific expertise and are assigned to conduct independent assessments of proposed projects before granting approvals. Instead of conducting such assessments, these agencies often endorse the decision of an apex environmental agency such as SEIAA or MoEFCC (Hegde et al., 2020). The clearance letter issued by the KSCZMA acknowledges the original EC issued by the SEIAA but does not make any reference to the complaints of fisherfolk about the cumulative impacts on the ecologically sensitive zone. Moreover, the trivial conditions laid out in the letter, such as no dumping allowed into the sea of sewage from the construction activities, raise serious doubt about the KSCZMA's interest in objectively assessing the environmental risk of the project.

The fishermen association, as noted earlier, managed to obtain a temporary stay order from the Karnataka High Court to immediately stop construction activities. In the court proceedings, the burden of proof for challenging the information reported in the EIA report rested on the local community. Furthermore, Indian courts often give serious consideration to the assessments and advice offered by scientific committees and competent regulatory agencies in adjudicating environmental disputes. If agencies like the KSCZMA and KSPCB compromise the objectivity, rigour and purpose of their work, they undermine their reputation as impartial interpreters of the scientific information used in the courts. And in this case, their actions discouraged the Karwar fisherfolk and citizens who were seeking remedial environmental justice. As India pursues its dream for port-led prosperity under the Sagarmala program, the role of Coastal Regulation Zone authorities and other scientific institutions as neutral arbiters of scientific knowledge becomes even more pivotal than in the past. If they fail, the price will ultimately be paid by coastal environments and the often-marginalised communities dependent on those environments for their livelihoods.

References

Agrawal, D.K., Lodhi, M.S. and Panwar, S. (2010) "Are EIA studies sufficient for projected hydropower development in the India Himalayan region?" *Current Science*, 98(2), 154–161.

Aserkar, R. (2005) "Indian ports: Gateways to economic development," *Foreign Trade Review*, 39(4), 24–40.

Balaji, B. (2018) *Challenges towards Sustainable Port Development in India: The Adverse Effects of Port Development on Coastal Ecology and Community in Ennore: A Case Study.* Dissertation, World Maritime University, 672. Available at: https://commons.wmu.se/all_dissertations/672 (Accessed: 17 December 2020).

Bandekar, P., Neelkantan, K. and Kakati, V.S. (2011) "Biodiversity of crabs in Karwar mangrove environment, west coast of India," *Recent Research in Science and Technology*, 3(4), 1–5.

Banham, W. and Brew, D. (1996) "A review of the development of environmental impact assessment in India," *Project Appraisal*, 11, 195–202.

Bhaitkol Bandaru Nirashritara Yantrikruta Dhoni Meenugarara Sahakari Sangha v Karnataka Maritime Board, W.P. 1332/2020, Karnataka High Court, Bangalore.

Boer, W., Slinger, J.H., Kangeri, A., Vreugdenhil, H., Taneja, P., Appeaning Addo, K. and Vellinga, T. (2019) "Identifying ecosystem-based alternatives for the design of a seaport's marine infrastructure: The case of Tema Port expansion in Ghana," *Sustainability*, 11, 6633.

CPR–Namati (2017) *Paralegals for Environmental Justice.* Version 2.0. New Delhi: Centre for Policy Research–Namati Environmental Justice Program. Available at: https://cprindia.org/system/tdf/policy-briefs/NAMATI%20PPG%20FULL_E-Version.pdf?file=1&type=node&id=6593 (Accessed: 17 December 2020).

CRZ (Coastal Regulation Zone) (2011) *Costal Regulation Zone Notification S.O. 19(E). Gazette of India, Extraordinary, Part-II, Section 3, Sub-section (ii) 6 January 2011.* Ministry of Environment, Forest and Climate Change. Available at: https://parivesh.nic.in/writereaddata/ENV/crz23.PDF (Accessed: 17 December 2020).

Damle, H. (2017) *Financial Analysis of Blue Economy, Sagarmala Case in Point.* PFPAC Report, New Delhi.

Deccan Herald (2020) "Fishermen protest Sagarmala project, call bandh," *Deccan Herald*, 13 January. Available at: https://www.deccanherald.com/state/karnataka-districts/fishermen-protest-sagarmala-project-call-bandh-794266.html (Accessed: 17 December 2020).

Department of Economic Affairs (2020) *Public Private Partnership in India.* PPP Cell Infrastructure Division, Government of India. Available at: https://www.pppinindia.gov.in/overview (Accessed: 17 December 2020).

Dilay, A., Diduck, A.P. and Patel, K. (2020) "Environmental justice in India: A case study of environmental impact assessment, community engagement and public interest litigation," *Impact Assessment and Project Appraisal*, 38(1), 16–27.

Directorate of Ports and Inland Water Transport (2017) Executive Summary, *Proposed IInd Stage Development of Commercial Karwar Port*. Environment Impact Assessment Report, Port Authority, Karwar division. Available at: https://kspcb.gov.in/PH/EXE-SUM-ENG_develop__karwar%20port_17012017.pdf(Accessed: 17 December 2020).

Down to Earth (2018) "Eco-sensitive regions could see flurry of construction activity as draft CRZ notification suggests," *Down to Earth*, 20 April. Available at: https://www.downtoearth.org.in/news/environment/eco-sensitive-regions-could-see-flurry-of-construction-activity-as-draft-crz-notification-suggests-60257 (Accessed: 17 December 2020).

Dwarakish, G.S. and Salim, A.M. (2015) "Review of role of ports in the development of the nation," *Aquatic Procedia*, 4, 295–301.

EIA Notification (2006) *Environment Impact Assessment Notification 2006 S.O. 1533*. Ministry of Environment and Forest, New Delhi.

Equations (2000) *Karnataka Coast: A Case for Better Protection*. Equitable Tourism Options, Bangalore. Available at: https://www.scribd.com/document/34573308/Karnataka-Coast-A-Case-for-Better-Protection (Accessed: 17 December 2020).

Gill, G.N. (2017) *Environmental Justice in India: The National Green Tribunal*. Abingdon, Oxon: Routledge.

Grumbine, R.E. and Pandit, M.K. (2013) "Threats from India's Himalaya dams," *Science*, 339(6115), 36–37.

G. Sundarrajan v Union of India and Others, Appeal No. 60 of 2018 (National Green Tribunal), Southern Zone Bench, Chennai.

Government of Karnataka (2019) *Infrastructure, Development, Ports and Inland Water Transport Department*. Available at: https://idd.karnataka.gov.in/english (Accessed: 17 December 2020).

Hanumagond, P.T. and Mitra, D. (2007) "Dynamics of Karwar coast, India, with special reference to study of tectonics and coastal evolution using remote sensing data," *Journal of Coastal Research*, 50, 842–847.

Haragi, S.B. and Naik, U.G. (2012) "Resilient intertidal biodiversity profile of Majali coast, Karwar, west coast of India," *Cibtech Journal of Zoology*, 1(1), 68–78.

Heaver, T.D. (1993) "The many facets of maritime economics, in association?" *Maritime Policy and Management*, 20(2), 121–132.

Hegde, M. and Dinesh, K. (2016) "Tadadi port project: In troubled waters," *India Legal*, 21 October. Available at: https://www.indialegallive.com/cover-story-articles/il-feature-news/karnatakas-tadri-port-project-under-the-scanner/ (Accessed: 17 December 2020).

Hegde, M., Patel, K. and Diduck, A.P. (2021) "Environmental clearance conditions in impact assessment in India: Moving beyond greenwash," *Journal of Environmental Assessment Policy and Management* (under review).

Huralimath, A. (2019) "Centre asks Karnataka government to reconsider clearance for expansion of Karwar port," *New Indian Express*, 29 August. Available at: https://www.newindianexpress.com/states/karnataka/2019/aug/29/centre-asks-karnataka-government-to-reconsider-clearance-for-expansion-of-karwar-port-2026118.html (Accessed: 17 December 2020).

Kaladharan, P., Zacharia, P.U. and Vijayakumaran, K. (2011) "Coastal and marine floral biodiversity along the Karnataka coast," *Journal of the Marine Biological Association of India*, 53(1), 121–129.

Krishna Vithoba Javekar v State of Karnataka and Others, W.P. 100981/0983/2018, Karnataka High Court, Dharwad.

KSCZMA (Karnataka State Coastal Zone Management Authority) (2018) *Coastal Zone Management Plan of Karnataka*, 7 August. Available at: https://karunadu.karnataka.gov. in/ksczma/english/Pages/CZMP.aspx (Accessed: 17 December 2020).

KSCZMA (Karnataka State Coastal Zone Management Authority) (2019) *Coastal Regulation Zone, Clearance Letter Granted by Karnataka State Coastal Zone Management Authority*, 30 December.

Menon, M. and Kohli, K. (2007) "Environmental decision-making: Whose agenda?" *Economic and Political Weekly*, 42, 2490–2494.

Motupalli, C. (2018) "Intergenerational justice, environmental law, and restorative justice," *Washington Journal of Environmental Law and Policy*, 8(2), 333–361.

Munim, Z.H. and Schramm, H.-J. (2018) "The impact of port infrastructure and logistics performance on economic growth: Mediating role of seaborne trade," *Journal of Shipping and Trade*, 1(3), 1–19.

Naik, D. and Kunte, P. (2016) "Impact of port structures on the shoreline of Karnataka, West Coast, India," *International Journal of Advanced Remote Sensing and GIS*, 5(5), 1726–1746.

Naik, U.G., Rathod, J.L. and Bhat, U.G. (2005) "Temporo-spatial distribution of Meiobenthic fauna in River Kali," *Environment and Ecology*, 23(2), 254–258.

Nayak, S. (2000) "Critical issues in coastal zone management and role of remote sensing" in Sudarshana, R., Mitra, D., Mishra, A.K., Roy, P.S. and Rao, D.P. (eds.), *Subtle Issues in Coastal Management*. Dehradun: Indian Institute of Remote Sensing, 75–98.

Nayak, V.N. (2017) *Impact of Proposed Port Expansion on Coast of Karwar*. Research report submitted by Karnataka Regional Science Centre, Karwar.

Ng, A.K.Y. and Song, S. (2010) "The environmental impacts of pollutants generated by routine shipping operations on ports," *Ocean and Coastal Management*, 53, 301–311.

Pring, G. and Pring, C. (2009) *Greening Justice: Creating and Improving Environmental Courts and Tribunals*. New Delhi: The Access Initiative.

Rajan, S.R. (2014) "Environmental justice in India," *Environmental Justice*, 7(5), 115–116. Available at: https://doi.org/10.1089/env.2014.7502.

Rathi, A.K.A. (2017) "Evaluation of project-level environmental impact assessment and SWOT analysis of EIA process in India," *Environmental Impact Assessment Review*, 67, 31–39.

Schlosberg, D. (2004) "Reconceiving environmental justice: Global movements and political theories," *Environmental Politics*, 13(3), 517–540.

SEIAA (State Environment Impact Assessment Authority) (2019) *Environmental Clearance Granted The Directorate of Ports and Inland Water Transport 23*. Available at: http://environmentclearance.nic.in/writereaddata/FormB/EC/EC_Letter/ 022020190QIRT4U48IND2017-EC.pdf (Accessed: 18 December 2020).

Shaikh, N., Rathod, J.L. and Durgekar, R. (2017) "Zooplankton diversity in river Kali, Karwar, west coast of India," *International Journal of Engineering Development and Research*, 5(3), 495–500.

Shan, J., Yu, M. and Lee, C.-Y. (2014) "An empirical investigation of the seaport's economic impact: Evidence from major ports in China," *Transportation Research Part E: Logistics and Transportation Review*, 69(C), 41–53.

Shirodkar, P.V., Pradhan, U.K. and Vethamony, P. (2010) "Impact of water quality changes on harbour environment due to port activities along the west coast of India." Second International Conference on Coastal Zone Engineering and Management (Arabian Coast 2010), 1–3 November, Muscat, Oman.

Sivadas, S., Gregory, A. and Ingole, B. (2008) "How vulnerable is Indian coast to oil spills? Impact of MV Ocean Seraya oil spill," *Current Science*, 95(4), 504–512.

Sleeper, D.M. (2012) "Port significance: Contributions to competitiveness in Latin America and Asia," *Journal for Global Business and Community*, 3(1), 22–28.

Tenjing, Y.S., Deepak Samuel, V., Meenakshi, V.K., Siva Sankar, R. and Narasimhaiah, N. (2018) "Intertidal biodiversity of Karnataka, south west coast of India" in *Perspectives on Biodiversity of India*, vol. 3, 186–192. Fourth Indian Biodiversity Congress (IBC 2017) 10–12 March, Puducherry.

Vaidya, N.G. (2011) "*Stranded longbeakcommon dolphin, Delphinus capensis (gray, 1828) at Karwar, Karnataka*" in *Marine Fisheries Information Service, Technical and Extension Series*, 210,17.

Vining, A.R. and Boardman, A.E. (2008) "The potential role of public-private partnerships in the upgrade of port infrastructure: Normative and positive considerations," *Maritime Policy and Management*, 35(6), 551–569.

Vivekanandan, E. and Jeyabaskaran, R. (2012) *Marine Mammal Species of India*. Kochi: Central Marine Fisheries Research Institute.

Wang, T.-F. and Cullinane, K. (2006) "The efficiency of European container terminals and implications for supply chain management," *Maritime Economics and Logistics*, 8(1), 82–99.

Williams, G. and Mawdsley, E. (2006) "Postcolonial environmental justice: Government and governance in India," *Geoforum*, 37(5), 660–670.

13 Advancing environmental justice

Lessons from the Thoothukudi Sterlite Copper case

Aruna Kumar Malik, Alan P. Diduck and Kirit Patel

Introduction

Development that addresses the needs of underserved communities in India requires innovative public policies, programmes and institutions. Over the last three decades, India has adopted several such initiatives in an attempt to offset the adverse impacts of industrial activities. These measures include the creation of environmental assessment and clearance (EC) processes and related enforcement and compliance mechanisms (Ghosh, 2019). This chapter analyses these initiatives and explores their environmental justice implications in relation to the establishment, operation and 2018 shutdown of the highly contentious Sterlite Copper smelter in Thoothukudi (formerly Tuticorin), Tamil Nadu (Nath, 2018; Sofia, 2018). Sterlite Copper is a unit of Vedanta Limited, a large mining and metals company based in the United Kingdom. The Sterlite Copper plant in Thoothukudi has the capacity to produce 438,000 t of copper per annum, or 1,200 t/day. Before setting up the company in Thoothukudi, Sterlite had proposed it for the Ratnagiri District of Maharashtra in 1992. That proposal was accepted, construction commenced and the company invested Rs. 200 crores in the project. However, farmers protested against the project, believing that the smelting process would cause environmental pollution, and as a result, the District Collector directed the company to suspend its construction activity on 15 July 1993. The company then approached the states of Goa and Gujarat but ran into similar opposition from farmers and local communities. Ultimately, the plant site was proposed for Thoothukudi, a coastal town in Tamil Nadu in the Gulf of Mannar region (Figure 13.1) which abounds in biodiversity. About 2,000 marine species and another 200 terrestrial plant species are reported to have been found in the region. There are 21 islands near Thoothukudi, and these were declared a Marine National Park in 1986, with a view to protecting the unique and fragile flora and fauna in the region (GO Ms. No. 962, 10 September 1986). In addition, the perennial Tambararani River is an important source of water for cultivation in the region. The river is close to the plant site, making it vulnerable to pollution from the smelter. Furthermore,

DOI: 10.4324/9781003141228-13

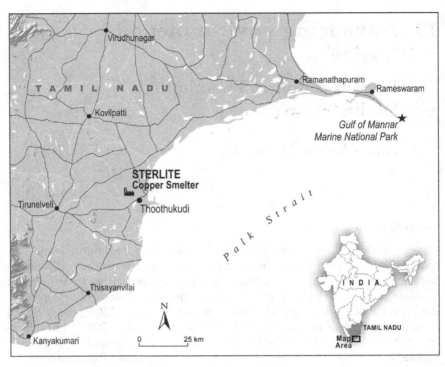

Figure 13.1 Map showing Thoothukudi city and the Gulf of Mannar Marine National Park.

village residents used groundwater as their source of drinking water, and this had been depleted considerably in recent years, creating distress for farmers, as well as domestic users (*Sterlite Industries (India) Limited v Tamil Nadu Pollution Control Board and Others*, 2013). Despite these problems, the Government of Tamil Nadu accepted the company's proposal and approved the setup of the copper smelter in Thoothukudi.

The struggle for justice

Early approvals and local opposition

On 27 January 1994, the Government of India, under rule 5(3)(a) of the Environment (Protection) Rules 1986, promulgated the Environment Impact Assessment Notification 1994, making ECs mandatory for expansion or modernization of any activity or for setting up new projects of the type listed in Schedule 1 of the notification. Section 2(I)(a) of the notification established the requirements and procedure for seeking ECs. Since the Sterlite plant was a project exceeding Rs. 50 crores, it was mandatory for the project proponent to obtain an EC from the Ministry of Environment and Forests (now the Ministry of Environment, Forest and Climate Change [MoEFCC]). However, an environmental impact assessment (EIA) was never conducted and associated public hearings were never held.

Despite the lack of an EIA and lack of an EC from the MoEFCC, state government approvals were granted. An EC was issued by Tamil Nadu's Environment and Forest Department on 16 January 1995, and on 22 May 1995, the Tamil Nadu Pollution Control Board (TNPCB) granted permissions under section 25 of the Water (Prevention and Control of Pollution) Act 1974 and section 21 of the Air (Prevention and Control of Pollution) Act 1981. The precise location approved for the plant was the State Industrial Progress Complex of Tamil Nadu (SIPCOT), Melavattan village, Thoothukudi Taluka, which was within 14 km of the Gulf of Mannar. Subsequently, the TNPCB issued a consent to operate on 14 October 1996, permitting the company to manufacture up to 234 t of blister copper per day and 638 t of sulphuric acid, despite the plant being near the Gulf of Mannar ecologically sensitive area.

Although Sterlite Copper commenced production on 1 January 1997, its EC and consents were challenged before the Madras High Court in 1996, 1997 and 1998 on grounds that the company had flouted environmental laws and regulations (explained further in the next section). Despite this litigation, in 1999, the TNPCB granted consent for production of two more products, namely phosphoric acid and hydrofluorosilicic acid. Furthermore, copper anode production limits were subsequently increased to 1,200 t/day, with approval of the MoEFCC and the TNPCB.

Local people complained about this expansion as well as ongoing operations, in an effort to prevent pollution and other environmental damage. Furthermore, the local residents believed there was a gas leak in the plant that caused air and water pollution, resulting in human health impacts such as eye, skin and lung irritation. The residents launched a complaint to the TNPCB and the District Collector, but both authorities found no evidence of pollution and issued a clean chit to the company. At the same time, the company was monitored owing to various interim orders issued by environmental regulatory authorities.

Some of the staff of the company who were local residents aimed to prevent pollution and make the SIPCOT premises as green as possible. Despite these efforts, an incident occurred on 20 August 1997 in which a sulphur dioxide gas leak from one of the Sterlite smelters caused staff of the Tamil Nadu Electricity Board's substation (a plant located across from the Sterlite unit) to complain of headaches, coughing and throat irritation. A similar incident was reported by staff members of the All India Radio Station located near Sterlite. In this case, 11 people were hospitalized for inhalation of noxious gas. In addition, there were unreported incidents that occurred from time to time in the villages surrounding the copper plant. It is evident that the presence of intensive air pollution caused deterioration of health to residents in and around the copper smelter plant (Bapat, 2005; Sangomla, 2018).

Court challenges

The Madras High Court ordered closure of the plant on 23 November 1998, on the basis of a petition filed by a non-governmental organization called the National Trust for Clean Environment. The petition stated that the EC granted by the state government was in violation of the provisions of the Environment (Protection) Act 1986, section 21 of the Air (Prevention and Control of Pollution) Act 1981 and section 25 of the Water (Prevention and Control of Pollution) Act 1974. After the first hearing, the court directed the National Environmental Engineering Research Institution (NEERI) to carry out a study on the environmental impacts of the plant and submit its findings to the court (*National Trust for Clean Environment v Union of India and Others*, 2010).

NEERI found that Sterlite had flouted various environmental regulations, including the requirement to conduct an EIA [NEERI (National Environmental Engineering Research Institute), 1998]. The report also indicated that the ECs and consents obtained from the central and state governments were in contravention of relevant statutory requirements, such as (1) allowing the company to establish operations within a 25-km radius of an ecologically sensitive area in the Gulf of Mannar; (2) preparing an inadequate EIA report based on only a month's worth of data; and (3) relaxing green belt requirements without adequate justifications. The report indicated that Sterlite had failed to develop the green belt and was producing chemicals not authorized under its clearance from the TNPCB. Furthermore, the plant had contaminated the groundwater with arsenic, lead, selenium, aluminium and copper. Additionally, it had been located within 14 km of islands in the Gulf of Mannar, thereby violating the No Objection Certificate the company had received from the MoEFCC, which included a condition to keep 25 km away from the Gulf of Mannar.

Despite objections of the central government, TNPCB and Sterlite to NEERI's research findings during a public interest suit against the company, the High Court ordered closure of the plant in 1998, citing violations of various environmental statutes and Sterlite's licence. The High Court also declared that employees of the plant would be entitled to compensation under section 25-FFF of the Industrial Disputes Act 1947. Furthermore, it directed the District Collector of Thoothukudi to take all necessary steps for re-employment of the workforce so as to protect their livelihoods, having due consideration for their educational qualifications and technical expertise (Madras High Court, 2018).

Sterlite appealed this order to the Supreme Court of India. Subsequently, reconsidering the consequences of its order, the Madras High Court revoked its order of closure on 23 December 1998 and temporarily allowed the plant to resume operations on an experimental basis from 26 December 1998 to 28 February 1999. The company also challenged this order in the apex court (*National Trust for Clean Environment v Union of India and Others*, 2010), writ petition Nos 15501–15503 of 1996, arguing that its plant had complied with the terms and conditions set forth by the TNPCB and petitioning for permission to operate at full capacity. In the meantime, within a month – on 2 March 1999 – the incident affecting the All India Radio Station staff (noted above) occurred. While the matter was pending before the Supreme Court, the company successfully petitioned the Madras High Court for permission to start production. The court ordered the TNPCB to allow production on a temporary basis. Paying no heed to the radio station incident, the TNPCB granted permission to the company to nearly double its production from 40,000 to 70,000 t/annum (Madras High Court, 2010). In response, local people protested in an effort to stop production because the company was in violation of several regulatory requirements, such as establishing the plant within an ecologically sensitive area and failing to develop a green belt.

After a series of further violations by the company of both legislation and its operating licence, the Tamil Nadu government in 2004 ordered the plant to be shut down [TNPCB (Tamil Nadu Pollution Control Board), 2018]. The company challenged the order and sought relief from the Supreme Court. As a result, the court constituted the Supreme Court Monitoring Committee to inspect the plant and its waste management mechanisms (*Sterlite Industries (India) Limited and Others v Union of India and Others*, 2010). The committee's inspection team found Sterlite's operations to be lacking and recommended that the company be denied a new EC to expand copper production from 391 to 900 t (*Sterlite Industries (India) Limited v Tamil Nadu Pollution Control Board and Others*, 2013). It also found that a number of associated sub-plants that were listed as part of the proposed expansion had already been built. The committee directed the TNPCB to inspect and take suitable action if the company had indeed constructed unlicensed production units. Within a day of the committee's inspection, the MoEFCC issued an EC to Sterlite for plants it had already begun to construct (*National Trust for Clean Environment v Union of India and Others*, 2010).

Later that year, the TNPCB submitted its report confirming that the company was engaged in unlicensed production (Maharajan and Samual, 2010). It found that an entire factory complex consisting of the copper smelter, a refinery, a sulphuric acid plant, a phosphoric acid plant, converters and a continuous cast rod plant were in varying stages of completion. The sulphuric acid plant had been completed in 2004 without having an EC. Additionally, none of these newly constructed plants had a construction licence from the TNPCB. Therefore, the Madras High Court ordered closure of the plant; however, the Supreme Court ruled on the matter, specifying that the company had maintained the standard established by TNPCB under the Water and Air Acts. In an interim order, the Supreme Court stayed the order from the Division Bench of the Madras High Court and, thus, allowed the company to continue production of copper, pending a final judgement (Supreme Court of India, 2010). The apex court's final judgement, pronounced in April 2013, accepted that Sterlite had polluted Thoothukudi and violated environmental laws. The court, however, did not order a shutdown because of copper's importance to national defence, the electrical equipment industry and infrastructure development, and because closure of the plant would affect jobs and the economy of the Tamil Nadu. The court fined Sterlite Rs. 100 crore and directed the district administration to use the interest amount for local infrastructure development under the supervision of the District Collector (*Sterlite Industries (India) Limited v Tamil Nadu Pollution Control Board and Others*, 2013).

In the same year, the TNPCB received complaints about throat and eye irritation, suffocation and skin-related problems from New Colony and Keela Shunmuga Puram villages and other areas of Thoothukudi town, which are around 6 and 9 km away from the plant. These problems arose due to gas emissions, and local people alleged that the company was polluting the air by releasing SO_2 (sulphur dioxide). In fact, ambient air concentrations of SO_2 increased from 20 to 62 $\mu g/m^3$ in 2013, in contravention of the provisions of section 21 of the Air (Prevention and Control of Pollution) Act 1981 (*Sterlite Industries (India) Limited v Tamil Nadu Pollution Control Board and Others*, 2013). After receiving these complaints, the TNPCB served a show-cause notice on the company for contravening the conditions imposed in the consent issued under section 21. The company was asked to show why directions under section 31A of the act should not be issued for closure of the unit and disconnection of its power and water supply. The company's explanation did not satisfy the TNPCB, which directed closure of the plant with immediate effect (TNPCB order passed dated 29 March 2013). Subsequently, the company challenged the order before the National Green Tribunal (NGT) on 8 August 2013 (*Sterlite Industries (India) Limited v Tamil Nadu Pollution Control Board and Others*, 2013), primarily on the grounds that the closure order was arbitrary and discriminatory and failed to consider relevant materials. The company also argued that it was not given an opportunity to be heard. It stated that the order was based on irrelevant materials and unscientific data and ignored substantial facts of the incident. The company also claimed that the sulphuric acid plant bed was maintained at the required temperature using furnace oil, and the emissions were routed through a tail gas scrubber. The NGT put special emphasis on gas emissions and complaints related to health issues; however, it could not find substantial evidence of health problems among residents. Ultimately, the TNPCB's failure to produce proper records led the NGT to revoke the closure order, allowing the company to continue production (*Vedanta Limited v State of Tamil Nadu and Others*, 2018).

Public protests

Despite written complaints from local residents, the courts and environmental regulatory authorities, time and again, allowed Sterlite to continue in operation. As a result, public opposition against the plant continued to grow. However, these protests, which involved a broad cross section of society, petered out owing to differences of opinion among the movement's members. Initially, the Evangelical Church (Venkatesan, 2018) took the initiative in organizing protests involving a large number of fisherfolk. "It was natural that the fishermen were apprehensive regarding discharge of effluents into the ecologically fragile Gulf of Mannar, they feared that pollution would destroy fish breeding spots and jeopardise their livelihood" (Rajasekaran, 2018). These protests received little response from Sterlite and the district administration. The protests then intensified and developed into an anti-Sterlite movement. On 18 July 2017, the movement organized a massive human chain, with nearly 4,000 people participating, to pressure the government and the TNPCB to withdraw their consent orders for Sterlite Phase II (Shankar, 2021). This struggle against the expansion activities further intensified in 2018, with lack of enforcement of statutory provisions becoming a key bone of contention.

In January 2018, when the company started site preparation as part of their plans to double the capacity of the smelter, the local people became even more concerned for their health and started a protest in the small village of Kumarredyapuram. The first day of protest was 12 February 2018, after an attempt to meet the District Collector did not yield positive results. From 13 February until 23 March 2018, villagers from Kumarredyapuram sat in groups under a neem tree as a mark of protest against the plant (Alagu, 2018).

On 24 March 2018, the protest entered its 41st day and more than 100,000 residents of Thoothukudi flooded the streets, demanding an immediate closure of Sterlite's operations. The influential merchant association's call to its members to shutter their shops for a day was the trigger. Artisanal fisherfolk, shank drivers, small salt pan manufacturers, the Tuticorin Chamber of Commerce, auto rickshaw unions, minibus drivers and tea stall vendors quickly joined the call and stayed off work (Scroll, 2018). They called for an immediate halt of the work to construct a new copper smelter complex in Therku Veerapandiapuram, a suburban locality west of Thoothukudi town, and closure of the existing plant. A public meeting at Thoothukudi's VVD junction, under the slogan "Ban Sterlite, Save Thoothukudi," was energized by children shouting slogans, considerable youth energy and colourful dances by environmental activists. The police observed the meeting and were outnumbered by the protesters, but no untoward incidents occurred. Between 25 March and 20 May, eight to ten more villages joined the movement and put up makeshift tents in Thoothukudi (Scroll, 2018).

On 22 May 2018, the 100th day of protest, a rally demanding the plant's closure turned violent, with 13 people killed in police fire (Economic and Political Weekly, 2018). The next day, the police opened fire on protesters who were pelting stones at police, and this claimed the life of a youth. The protests then accelerated, with the number of protesters swelling as each day passed (Radhakrishnan, 2018). On 28 May, the Madurai Bench of the Madras High Court passed an order staying construction of Sterlite Phase II. The next day, the TNPCB disconnected the plant's electricity and water supply (BBC, 2018). Also, the District Collector of Thoothukudi sealed the plant and announced its permanent closure. After less than a month, on 26 June 2018, the company challenged these orders before the NGT and sought permission to operate the plants. The NGT established a three-member committee to probe the closure of the copper plant and examine its impacts on groundwater quality, air quality and overall environmental sustainability. On the basis of the committee's report, the NGT set aside the TNPCB's and District Collector's closure orders and on 15 December 2018 directed the TNPCB to restore electricity to the plant. In its final order, the NGT stated that the grounds on which the government took the decision to shut the plant did not justify the closure orders; the plant was, therefore, permitted to restart production (The Wire, 2018).

On 2 January 2019, the Tamil Nadu government and the TNPCB filed an appeal before the Supreme Court, challenging the orders of the NGT. The TNPCB alleged that Sterlite had not furnished the groundwater analysis report, had failed to remove copper slag stored around the Upper River and had not analysed heavy metals in the ambient air around the unit. On 8 January 2019, the Supreme Court ruled that the NGT did not have jurisdiction to order reopening of the plant. The court also refused to overrule the Madras High Court, which had stayed the NGT's order. In doing so, the court directed Sterlite to take its plea for reopening back to the High Court (Bharani and Mazumdar, 2018).

Key environmental justice principles and their constitutional context

Principles

"Each society has its laws, norms and values, some of which distribute the benefits and burdens resulting from social and economic development. Principles of distributive justice can be thought of as providing moral guidance for the institutional processes and structures that affect the distribution of burdens and benefits in society" (Lamont and Favor, 2017). Distributive justice is, of course, a key principle of environmental justice, concerning itself with the distribution of environmental "goods" (e.g., access to green space) and environmental "bads" (e.g., pollution and risk factors) among different social groups and the fairness or equity of this distribution (Raphael, 2019; Schlosberg, 2004). Equity in distribution requires adherence to the other three key principles of environmental justice examined in this book.

It is nearly impossible to achieve distributive equity without paying careful attention to recognitional justice, that is, recognizing the varying communities, values, interests, needs, aspirations and intersections in society (Malin, Ryder and Lyra, 2019; Massarella, Sallu and Ensor, 2020; Whyte, 2018). In turn, heeding recognitional justice provides a foundation for procedural justice, which itself is a force for distributive equity. Procedural justice entails providing meaningful opportunities for affected communities to engage in environmental governance, including formal planning, assessment and approval processes. It also involves access to the courts in order to protect rights and pursue remedies. Underlying procedural justice is timely access to complete and accurate information about: potential environmental costs and benefits of a proposed project or decision; the applicable regulatory framework; gaining access to the courts; and possible legal remedies that are available (Dilay, Diduck and Patel, 2020; Pring and Pring, 2009; Rajamani, 2007; United Nations Economic Commission for Europe, 2019). Finally, restorative justice, a principle borrowed from criminology (e.g., Armour, 2012), is concerned with the extent to which negative environmental and social impacts of a project or decision are mitigated or remedied (Motupalli, 2018; Raphael, 2019). Careful attention to restorative justice can help guide environmental planning, assessment and approval processes and remedial decisions made by the courts.

Constitutional context

Important elements of the administration of environmental justice in India, incorporated in legal documents and judicial decisions, derive their authority from basic principles enshrined in the Constitution. A cluster of environmental legal instruments, governmental authorities, policies, programmes and case law (including Supreme Court decisions) have been developed to protect and improve the environment and guarantee the right to a healthy environment.

Article 48-A of the Constitution provides that the state "shall endeavour to protect and improve the environment and to safeguard the forests and wildlife of the country." This provision is under Part IV of the Constitution, which prescribes the Directive Principles of State Policy. These principles do not have the same enforceable constitutional authority as the Fundamental Rights, found in Part III of the Constitution. However, in 1987 the Supreme Court of India suggested that there could be an implicit right to a healthy environment among the Fundamental Rights. The Supreme Court mentioned in its judgement that people have a right "to live in a healthy environment with minimal disturbance of the ecological balance" (*Rural Litigation and Entitlement Kendra v Uttar Pradesh and Others*, 1987). Similarly, the High Court of Andhra Pradesh stated that "slow poisoning caused by environmental pollution and spoliation should be treated as amounting to a violation of article 21 of the constitution" (*T. Damodar Rao and Others v Special Officer, Municipal Corporation of Hyderabad*, 1987).

Later, in 1991, the Supreme Court clarified that the right to life is a fundamental right under Article 21 of the Constitution. From this, the argument has been made that the right to have a clean environment congenial to human existence is a part of the right to life. Article 21 guarantees "protection of personal liberty and life" (Bakshi, 2011), and life encompasses within its ambit the protection and preservation of the environment, ecological balance, clean drinking water and pollution-free air.

The state, thus, has a duty under Article 21 to forge policies that maintain ecological balance and a hygienic environment (Bakshi, 2011). Moreover, anything that endangers or impairs the quality of life of citizens in violation or in derogation of laws must be viewed seriously. Articles 39, 47 and 48-A of the Constitution by themselves and collectively cast a duty on the state to secure the health of the people and improve and protect the environment. If anything endangers or impairs that quality of life in derogation of laws, a citizen has the right of recourse to Article 32 of the Constitution for preventing the pollution of water or air which may be detrimental to the quality of life (*Subhash Kumar v State of Bihar*, 1991).

Discussion: environmental justice denied

The Sterlite plant has been called a "negation of environmental justice" and a reminder of the stranglehold industry has over the state in the ease of doing business era (Sahu, 2018; also see Chapter 3, this volume). This stranglehold is used to further commercial interests by ignoring and often violating environmental regulations and standards, but at the same time, it has fuelled social movements opposed to predatory industrial activity (Sahu, 2018).

In the Sterlite case, lack of compliance with the EIA notification undermined important basics of procedural justice: fairness and meaningful participation in decision-making. Local communities affected by the plant were not given a fair chance to be heard about the impact of the plant, a direct violation of procedural fairness. Furthermore, by not completing an EIA, the government and project proponent failed to perform their due diligence and undermined the possibility of advancing distributive justice. Environmental regulatory authorities did impose retributive justice by ordering shutdown of the plant from time to time as a principle of fairness in the punishment of wrongs, but that would have been cold comfort to the local residents who had already been harmed by the plant. The procedural and distributive injustices in this case were aided by the central government's repeal of the requirement for proponents of industrial projects to obtain consents from locally affected Gram Sabhas (usually led by village elders) in consultation with civil society members. This was considered a major step in economic reform and benefitted many industries and factories, including the Sterlite unit in Thoothukudi, but for local communities, it seriously undercut fundamental aspects of environmental justice.

In this context, we are reminded of a landmark event in the history of environmental movements in India started in 1977 by the then Ganjam (now Berhampur) panchayat in Odisha. The petition was submitted to the Petition Committee of Rajya Sabha against massive pollution caused by the production of chlorine gas by Jayashree Chemicals. This was the first ever petition submitted to the Parliament by a panchayat demanding environmental justice. The demand was not to close the industrial unit, but rather, it was to adopt measures for protecting air and water in the area, including for new legislation to help in this regard. As a result, the committee recommended that there should be legislation to protect air and water from pollution. The central government later enacted a series of laws, but because of a lack of enforcement, people still complain that the air and water of the region are contaminated (Iyer, 1992; Mishra and Tripathy, 1978; Sahu, 2018).

Nevertheless, this early case confirms the valuable role that can be played by environmental regulatory institutions in advancing environmental justice, and the final result stands in contrast to the legal shortcomings and inconsistencies seen in the Sterlite case. Judgements have been delivered by not less than three courts (the Supreme Court, Madras High Court and NGT), and although the salient facts of the case remain unchanged, the substantive judgements vary considerably. In fact, as we saw above, environmental laws have changed over the years and many amendments have been passed, particularly with respect to EIA. The policy changes have in some ways improved environmental protection but, in many ways, have focused on implementing an economic development agenda. One of the most important legal instruments that could be used to advance environmental justice in India is the Environment Impact Assessment Notifications of 1994 and 2006 (Principal Rules). However, the notification has failed to be a meaningful tool for this purpose. Important drawbacks in the notification that can easily block distributive and restorative justice are the framing of EIA terms of reference by the project proponent, the exemptions given to certain projects and limitations of the appraisal process. The limited scope and ambition of the public engagement and hearing processes compound these problems and further act as a barrier to recognitional and procedural justice (see Chapters 6, 8 and 9, this volume).

In addition, the Sterlite case reveals how lack of enforcement of EC conditions and other regulatory requirements presents a formidable hurdle to achieving environmental justice. Despite several closure notices issued by the TNPCB and Madras High Court, the Sterlite Copper plant continued to operate until 2018. Sterlite was consistently able to use the courts to out-manoeuvre the enforcement efforts of central and state regulatory agencies. This suggests a strong need for increased capacity and perhaps will on the part of both levels of government to carefully monitor and rigorously enforce regulatory requirements. An avenue for increasing monitoring capacity is to institute a system of community-based monitoring or citizen science (e.g., Moyer, Fitzpatrick, Diduck and Froese, 2008; Sharpe and Conrad, 2006; Steven et al., 2019) that draws on India's extensive experiences with social audit systems, such as the one established by the Mahatma Gandhi National Rural Employment Guarantee Act 2005. Such a system could monitor the environmental and social impacts of a project and provide current data on an ongoing basis to regulatory agencies (Courville, Parker and Watchirs, 2003). It could also, if designed well, recognize and value the different voices of stakeholders, including marginalized and poor groups, whose interests are rarely considered.

Conclusion

Advancing environmental justice is an important element of good governance and sustainable development. However, EC, enforcement and compliance mechanisms failed to bring environmental justice in the Sterlite Copper plant case. Few, if any, aspects of environmental justice – procedural, distributive, recognitional or restorative – were well served by the regulatory, judicial and other administrative institutions in this case. Although the copper plant was ultimately closed (at the time of writing in 2020), the enduring legacy in this case is the serious harm suffered by local ecosystems and people, and the police action that led to the deaths of protesters in May 2018 (Economic and Political Weekly, 2018).

India's Constitution provides remedies for violation of the fundamental rights of citizens. But the judicial system is generally not accessible to laypeople due to inherent drawbacks in the system. Furthermore, environmental law has failed to incorporate necessary provisions to address environmental justice. One of the solutions developed by the judiciary is the liberalization of locus standi and the promotion of public interest litigation and social action. Despite its shortcomings (e.g., Karim, Vincents and Rahim, 2012), public interest litigation has contributed to advancing environmental justice for laypeople through the actions of public-spirited citizens. The Supreme Court has, thus, provided scope and judicial remedies to treat the right to a healthy environment as fundamental and capable of being protected by citizens and civil society by means of writ petitions. Moreover, the constitutional right to a healthy environment has contributed to improvements in the recognition of procedural justice, including access to information and participation in decision-making. In spite of these advancements, environmental degradation is widespread and the consequences have been unevenly distributed among the people, in violation of the principles of equality enshrined under Article 14 of the Constitution. The inequitable treatment of people in the administration of environmental law results in the denial of environmental justice. Environmental legal frameworks in India have failed not only by inadequately addressing environmental issues but also by creating barriers to environmental justice for marginalized sections of society.

The Sterlite case shows the enormous costs to parties seeking environmental justice, including loss of time, energy, capital, social cohesion, livelihoods and in some cases life itself. The case reveals that it is high time to bring additional innovative solutions to complex environmental issues and legal battles such as those seen in the Sterlite case. Serious gaps persist in environmental governance. Environmental impact assessment and clearance processes need more inclusive and transparent public engagement mechanisms. Pollution control boards and other regulatory agencies need additional capacity when it comes to monitoring and compliance enforcement. With respect to this last issue, community-based monitoring following a social audit model offers a promising way forward.

References

Alagu, K. (2018) "Sterlite Protest explainer: Recapping the what, why and how of Thoothukudi's battle with a copper smelting plant," *New India Express*, 22 May. Available at: https://www.newindianexpress.com/states/tamil-nadu/2018/may/22/sterlite-protest-explainer-recapping-the-what-why-and-how-of-thoothukudis-battle-with-a-copper-sm-1818122.html (Accessed: 8 December 2020).

Armour, M. (2012) "Restorative justice: Some facts and history," *Tikkun*, 27, 25–65.

Bakshi, P. (2011) *Constitution of India*. New Delhi: Universal Law Publishing.

Bapat, J. (2005) *Development Projects and Critical Theory of Environment*. New Delhi: Sage Publications.

BBC (2018) "India shuts Tamil Nadu smelting plant after deadly protests," *BBC News*, 24 May. Available at: https://www.bbc.com/news/world-asia-india-44234715 (Accessed: 8 December 2020).

Bharani, V. and Mazumdar, R. (2018) "Tuticorin protest: Tamil Nadu government orders permanent closure of Sterlite plant," *The Economic Times*, 29 May. Available at: https://economictimes.indiatimes.com/news/politics-and-nation/tamil-nadu-government-orders-permanent-closure-of-sterlite-plant-in-tuticorin/articleshow/64355730.cms?from=mdr (Accessed: 16 November 2020).

Courville, S., Parker, C. and Watchirs, H. (2003) "Introduction: Auditing in regulatory perspective," *Law and Policy*, 25(3), 179–184.

Economic and Political Weekly (2018) "Why Thoothukudi exploded," *Economic and Political Weekly*, 53(21), 26 May.

Dilay, A., Diduck, A.P. and Patel, K. (2020) "Environmental justice in India: A case study of environmental impact assessment, community engagement and public interest litigation," *Impact Assessment and Project Appraisal*, 38(1), 16–27.

Ghosh, S. (2019) *Indian Environmental Law: Key Concepts and Principles*. Hyderabad: Orient BlackSwan.

Iyer, V.R.K. (1992) *Environmental Protection and Legal Defense*. New Delhi: Sterling Publishers.

Karim, S., Vincents, O.B. and Rahim, M.M. (2012) "Legal activism for ensuring environmental justice," *Asian Journal of Comparative Law*, 7(1), 1–44.

Lamont, J. and Favor, C. (2017) "Distributive justice" in Edward N. Zalta (ed.), *The Stanford Encyclopedia of Philosophy*, Winter 2017 edn. Available at: https://plato.stanford.edu/archives/win2017/entries/justice-distributive (Accessed: 16 November 2020).

Madras High Court (2010) W.P. (MD) No. 15501 to 15503 of 1996, in W.P. (MD) No. 5769 of 1997, in W.P. (MD) No. 16861 of 1998, in W.P. (MD) No. 21272, 21274, 21275, 21276, 21278 of 1996, in W.P. (MD) No. 9593 of 1997, 8044 to 8046 and 10274, 1999).

Madras High Court (2018) W.M.P. (MD) No. 10257 of 2018 in W.P. (MD) No.11220, 2018.

Maharajan, K. and Samual, K. (2010) "Impact of environmental pollution on health: A sociological study in Tuticorin industrial town, TN, India," *Journal of Ecobiotechnology*, 2(2), 51–56.

Malin, S.A., Ryder, S. and Lyra, M.G. (2019) "Environmental justice and natural resource extraction: Intersections of power, equity and access," *Environmental Sociology*, 5(2), 109–116.

Massarella, K., Sallu, S.M. and Ensor, J.E. (2020) "Reproducing injustice: Why recognition matters in conservation project evaluation," *Global Environmental Change*, 65, 102181.

Mishra, A. and Tripathy, S. (1978) *Chipko Movement: Uttarakhand Women's Bid to Save Forest Wealth*. New Delhi: Gandhi Peace Foundation.

Motupalli, C. (2018) "Intergenerational justice, environmental law, and restorative justice," *Washington Journal of Environmental Law and Policy*, 8(2), 333–361.

Moyer, J., Fitzpatrick, P., Diduck, A.P. and Froese, B. (2008) "Toward community-based monitoring in the hog industry in Manitoba," *Canadian Public Administration*, 51(4), 637–658.

Nath, A. (2018) "Tamil Nadu govt to rename more than 3000 places," *India Today*, 12 December. Available at: https://www.indiatoday.in/india/story/tamil-nadu-3000-places-rename-1408127-2018-12-12 (Accessed: 10 December 2020).

National Trust for Clean Environment v Union of India and Others (2010) Writ Petitions Nos. 15501 to 15503 of 1996 & 5769 of 1997 & 16861 of 1998 & W.M.P. Nos. 21272, 21274 to 21276, 21278 of 1996 & 9593 of 1997 & 8044 to 8046 & 10274 of 1999, Decided On 28 September 2010, Madras High Court.

NEERI (National Environmental Engineering Research Institute) (1998) *Evaluation on Pollution and Environmental Protection Systems at Sterlite Industries (India) Ltd., Tuticorin*. Nagpur.

Pring, G. and Pring, C. (2009) *Greening Justice: Creating and Improving Environmental Courts and Tribunals*. New Delhi: The Access Initiative.

Radhakrishnan, S. (2018) "The Hindu explains: Sterlite protests," *The Hindu*, 23 May. Available at: https://www.thehindu.com/news/national/tamil-nadu/the-hindu-explains-sterlite-protests/article23969542.ece (Accessed: 10 December 2020).

Rajamani, L. (2007) "Public interest environmental litigation in India: Exploring issues of access, participation, equity, effectiveness and sustainability," *Journal of Environmental Law*, 19(3), 293–321.

Rajasekaran, I. (2018) "Saga of a struggle," *Frontline*, 27 April. Available at: https://frontline.thehindu.com/the-nation/saga-of-a-struggle/article10106626.ece (Accessed: 10 December 2020).

Raphael, C. (2019) *Engaged Scholarship for Environmental Justice: A Guide*. Santa Clara, CA: Santa Clara University.

Rural Litigation and Entitlement Kendra v Uttar Pradesh and Others (1987) AIR 1987 SC 359.

Sahu, S.N. (2018) "Sterlite protest is a call for Swaraj," *Down to Earth*, 6 June. Available at: https://www.downtoearth.org.in/blog/environment/sterlite-protest-is-a-call-for-swaraj-60779 (Accessed: 12 December 2020).

Sangomla, A. (2018) "Sterlite typifies all that's wrong with environmental governance in India," *Down to Earth*, 27 June. Available at: https://www.downtoearth.org.in/news/governance/sterlite-typifies-all-that-s-wrong-with-environmental-governance-in-india-60877 (Accessed: 10 December 2020).

Schlosberg, D. (2004) "Reconceiving environmental justice: Global movements and political theories," *Environmental Politics*, 13(3), 517–540.

Scroll (2018) "Tamil Nadu: Thousands protest against Sterlite Copper's plan to expand its plant in Thoothukudi," *Scroll.in*, 25 March. Available at: https://scroll.in/latest/873229/tamil-nadu-thousands-protest-against-sterlite-coppers-plan-to-expand-its-plant-in-thoothukudi (Accessed: 16 November 2020).

Shankar, K. (2021) "How a retired Indian professor took on a mining giant – and won," *Aljazeera*, 7 January. Available at: https://www.aljazeera.com/features/2021/1/7/india-how-a-retired-professor-took-on-a-mining-giant-and-won (Accessed: 15 January 2021).

Sharpe, A. and Conrad, C. (2006) "Community based ecological monitoring in Nova Scotia: Challenges and opportunities," *Environmental Monitoring and Assessment*, 113, 395–409.

Sofia, L. (2018) "As Sterlite plant expands, a city erupts in protest," *The Wire*, 4 April. Available at: https://thewire.in/environment/anti-sterlite-protests-fuelled-by-weak-sanctions-and-political-connections (Accessed: 10 December 2020).

Sterlite Industries (India) Limited and Others v Union of India and Others (2010) CA Nos. 2776-2783 of 2013 (Arising out of SLP (C) Nos. 28116-28123 of 2010) (The Supreme Court of India).

Sterlite Industries (India) Limited v Tamil Nadu Pollution Control Board and Others (2013) Appeal No. 57 of 2013, [Appeal No. 22 of 2013 (SZ)] and Appeal No. 58 of 2013, [Appeal No. 23 of 2013 (SZ)] (National Green Tribunal, New Delhi).

Steven, R., Barnes, M., Garnett, S.T., O'Connor, J., Oliver, J.L., Robinson, C. and Fuller, R.A. (2019) "Aligning citizen science with best practice: Threatened species conservation in Australia," *Conservation Science and Practice*, 1(10), e100.

Subhash Kumar v State of Bihar (1991) AIR 420, SC (1) 598.

Supreme Court of India (2010) Interim Order, passed on 1 October 2010, staying the judgement of the Madras High Court.

T. Damodar Rao and Others v Special Officer, Municipal Corporation of Hyderabad (1987) AIR 1987, AP 171.

The Wire (2018) "Anti-Sterlite protest: 11 killed in police firing in Tamil Nadu," *The Wire*, 22 May. Available at: https://thewire.in/environment/vedanta-sterlite-copper-tuticorin (Accessed: 10 December 2020).

TNPCB (Tamil Nadu Pollution Control Board) (2018) Processing of the pollution control board No. TS1/TNPCB/FN217/TTN/RL/W&A/2018 dated 23/05/2018, Madhurai (GO (W&A) no 72.

United Nations Economic Commission for Europe (2019) *Convention on Access to Information, Public Participation in Decision-Making and Access to Justice in Environmental Matters*. Available at: https://www.unece.org/env/pp/contentofaarhus.html (Accessed: 14 January 2019).

Vedanta Limited v State of Tamil Nadu and Others (2018) Appeal No. 87/2018 (M.A No. 1741/2018 & M.A No. 1747/2018) (National Green Tribunal, Principal Bench, New Delhi).

Venkatesan, T.S. (2018) "Sterlite violence: Ominous nexus of church, jihadis, Naxals," *Organiser – Voice of the Nation*, 21 June. Available at: https://www.organiser.org/Encyc/2018/6/21/Sterlite-Violence-Nexus-of-Church-Jihadis-Naxals.html (Accessed: 11 December 2020).

Whyte, K. (2018) "The recognition paradigm of environmental injustice" in Holifield, R., Chakraborty, J. and Walker, G. (eds.), *The Routledge Handbook of Environmental Justice*. New York: Routledge, 113–123.

14 Judicial environmentalism

The thorny case of *Prosopis juliflora*

R. Seenivasan

Introduction: History of *Prosopis juliflora*

Prosopis juliflora (referred to as prosopis hereafter) is an exotic plant brought from Jamaica, initially introduced in South India in the 19th century. The introduction of prosopis is traced to the efforts of the Board of Revenue and the Forest Department during the colonial times under the Madras Presidency, which comprised the present-day states of Andhra Pradesh and Tamil Nadu and parts of Karnataka and Kerala. In the latter half of the 19th century, there were many reasons to search for a plant that would provide fuel and fodder and other economic uses. The use of animal dung as fuel by households (especially dung from cows, bullocks and buffaloes) was rampant, and this practice reduced the availability of manure for agriculture. J.A. Voelcker (1893), in his famous report on "improving Indian agriculture," dedicated a whole chapter on how to make available more wood and made recommendations to speed up ongoing efforts in the policies and practices of the revenue, agriculture and the forest departments of the government. He recommended developing "fuel and fodder reserves" and raising "plantations along canal banks and railway lines." Thus, it was during this time of great scarcity of wood and fodder that prosopis came into the country. The efforts of the Forest Department (Voelcker, 1893, 157–198) in experimenting with various options are noteworthy. The plant offered hard and valuable timber, shade, food for cattle and hedges.

Important to the prosopis story in South India is the prevalence of earthen-bunded reservoirs, or tanks, that were built centuries ago and remain a major part of the water infrastructure used by rural communities. The tank bunds are long and wide and at times have cart tracks or roads on top of them. The water is stored upstream of the bund and is released through sluices to the paddy fields located downstream. The bunds and sluices are formed in cascades and the networks of channels bring water into the tanks and also safely dispose of surplus water during the flood season. In 2010, Tamil Nadu had around 42,000 small and big tanks irrigating 0.5 million hectares of land; these were connected through hundreds of miles of feeder channels. The South Indian states have around 200,000 tanks in all, serving nearly 20 per cent

DOI: 10.4324/9781003141228-14

of the states' irrigated areas (Seenivasan, 2014). The semi-arid, tank-intensive and chronically drought-hit areas of the Rayalaseema region within the Madras Presidency were chosen as the ideal place for launching the introduction of prosopis. Kamalapuram in the present-day Kadapa District of Andhra Pradesh was said to be the first recipient and host of prosopis in South India. In 1876, R.H. Beddome, a notable among Indian foresters, sent a formal request to introduce the plant in the Presidency and thus received the first batch of seeds (Reddy, 1978). This introduction was done with a "public purpose" in mind and followed all scientific and policy measures. Since then, the introduction has not failed in its objectives and served the intentions of the government remarkably well.

A revived interest in promoting the plant was taken up in the 1960s, due to the widespread shortage of fuelwood in southern states. It was actively promoted in southern districts of Tamil Nadu, as alternative fuelwood (Narayanan, 2014). Dry and arid regions like Ramanathapuram were chosen, and seeds were extensively supplied to start plantations. A compendium on invasive species (CABI, 2019) describes multiple benefits derived from prosopis, especially by rural communities in arid regions across the world. The abundant availability of firewood for marginal and landless families, jobs from woodcutting and charcoal making are often-cited benefits from India's prosopis jungles (Pasiecnik et al., 2001; Walter and Armstrong, 2014). In addition, potential uses for medicines and food products for humans are also reported, although the extent of such uses is not well recognized (Malhotra and Misra, 1983; Tewari et al., 2013). Scientists from India's Central Arid Zone Research Institute still believe the reasons for introducing prosopis remain valid even in the present times. According to them, prosopis has offered firewood and fodder and helped conserve the soil (Edmond, 2017).

In 1978, a senior officer of the Andhra Pradesh Forest Department wrote an article commemorating the benefits prosopis had provided in the previous 100 years. The officer glorified the history and contribution of the plant and the role of the department in the following words:

> *Prosopis Juliflora* has crossed the continents and oceans, and has come to India at the invitation of the Foresters. She has served the poor and the Forest Department faithfully for years. If Teak has been the Royal Timber, Prosopis has been the Loyal Timber of the Poor. She completes a century of colonization, stabilization, and expansion in 1977, and let us celebrate the centenary of *Prosopis juliflora*, and sing her saga of achievements and glory.
>
> (Reddy, 1978, 17)

Incidentally, these accolades came at a time when tank farmers in the Rayalaseema region found it depressing to watch the plant colonizing their tank bunds and beds and canal and river banks, thereby reducing the irrigation performance of their tank systems (Reddy, 1978).

The spread of prosopis: Impacts and concerns

At present, prosopis is thriving in many parts of the country in open spaces like grasslands, tank beds and fallow agricultural lands. Though the spread is not uniform in intensity, the plant is common in most inland as well as coastal areas. The government has played a role in its spread through various promotional activities that introduced the plant for meeting different requirements (Pasiecnik et al., 2001).

The International Water Management Institute collected data from 50 irrigation tanks, using satellite and remote-sensing technology, in five tank-intensive districts of Tamil Nadu (Sakthivadivel, 2016). Their analysis indicates prosopis infestation in tank beds and channel systems varies between 30 and 50 per cent of the surface area. This level of spread results in several problems that affect the prospects of irrigated farming. It is also estimated, based on the above assessment, that 0.25 million hectares of tank beds that could otherwise be used for livestock grazing have become partly unusable because of the presence of the plant.

Above all, undertaking periodic tank rehabilitation and channel excavation becomes difficult and warrants additional expenses. Hence, the spread of prosopis leads to reduced water arrival into irrigation tanks and affects the productivity and production of crops from tank-fed areas. Furthermore, reduced storage levels in tanks lead to reduced recharging of groundwater, resulting in lowered water levels in drinking water wells. Thus, the spread of prosopis has severe implications for the state's irrigation, drinking water availability and livestock economy. Incidentally, the same research shows that infestations of prosopis on private lands also provide substantial benefits to landowners, such as providing firewood and wood for charcoal making and fodder (Sakthivadivel, 2016).

The usefulness of the plant is contested by various interest groups. Those who do not own land and are not directly connected with farming believe the existence of prosopis does not matter other than it offers some revenue without much effort. Charcoal making from prosopis is an occupation in many southern districts of Tamil Nadu and employs a considerable number of people. Charcoal from prosopis is an inexpensive energy source and is used by households and businesses. It is also used by some factories as an ingredient in rayon production. Therefore, many landowners who are not actively cultivating their small pieces of unirrigated land in arid regions willingly allow prosopis to thrive, as it provides considerable revenue for them after few years (Pasiecnik et al., 2001).

In general, threats of invasive plants are well documented, and many studies have raised serious concerns about the impacts of such plants on native environments and ecological processes (Baskin, 2002). This is true even in highly degraded places like mine wastelands, with some authors arguing that even there, invasive plants are dangerous to native flora and fauna over the long run (Singh, 2007). As far as irrigation tanks are concerned, the presence of thorns on prosopis plants makes it very difficult to desilt the channels and tank beds, and for the same reason, the cost involved in clearing the plant is high and unaffordable for many people (Palanisami and Balasubramanian, 1995; Seenivasan and Kumar, 2004; Vaidyanathan, 2001).

Over time, unmaintained channels no longer carry water into the tanks, and tank bunds become weak, leading to deterioration of tank networks. Thus, among many problems affecting the performance of irrigation tanks in South India, the infestation of prosopis has become an issue of prime importance.

In 1978, the Anantapur Zilla Parishad (the council of local bodies for Anantapur District) in Andhra Pradesh, where the plant was introduced 100 years earlier, had resolved to eradicate the plant in order to save the tanks and channel networks and asked for support from the state government. This effort was motivated by observing the decline in the number of tanks since the 1960s. For example, in Tamil Nadu, between 1960 and 2013, the tank-irrigated area decreased from 0.733 million hectares (40 per cent of total irrigation) to 0.38 million hectares (Government of Tamil Nadu, 2014). The status of tanks in Andhra Pradesh was even worse, and tank farmers therefore began an organized effort to remove prosopis. And it is not only South Indian farmers who have been concerned about prosopis; the call to eradicate the plant is widespread and has been circulating for at least the last four decades.

Though the concerns of tank farmers are genuine, blaming prosopis alone for tank problems is unfair. Throughout history, tanks have always suffered from poor repair and inadequate investment in upkeep (Mosse, 2003; Vaidyanathan, 2001, 2006). Prosopis does not survive in well-maintained tanks and channels that get filled every year, because waterlogged conditions prevent survival of the plant (CABI, 2019). Hence, the reason the plant thrives in tanks and channels has more to do with poor maintenance of water systems. Years of neglect and half-hearted tank development are one of the reasons the plants thrive. The growth of weeds and shrubs in fallow lands, including tank beds, channels and riverbeds, is a natural phenomenon. The growth of cactus and other thorny shrubs has also affected large areas throughout the 20th century and remains a menace. Costly tank rehabilitation projects have always been necessary, and until recently, special projects have often been undertaken to eliminate cactuses. In essence, the problems faced by the tanks due to native bushes and weeds were not very different from the present problem with prosopis.

The reason for the extensive spread on private lands is not difficult to find. Revenue from rainfed cultivation has been declining for a long time, and hence, farmers are better off by not undertaking cultivation in any serious manner. Therefore, fallowed lands that receive ruminants for grazing see invasions of prosopis. Over time, the plants grow into thick jungles and start offering benefits to the landowners in the form of wood. If landowners wanted to remove the plants, this would entail incurring substantial costs. The failure of the rainfed agriculture system in many parts of South India has in many ways been the fundamental cause of the spread of this plant. Even when the plants are removed, there is no guarantee they will not come back in the next season or a year later (CABI, 2019).

Thus, the challenge of permanent removal of prosopis from common as well as private lands has long been a subject of much interest among plant scientists. The use of human labour and mechanical devices such as excavators is necessary to remove the plant's branches and roots, although the use of herbicides such as 2,4-D has also been attempted by researchers and farmers. More importantly, even if prosopis is removed from an area, it can come back the next season unless efforts are made to remove every seedling that germinates in the area. Such a possibility is remote and may not happen in tanks that are mostly empty due to other reasons. Therefore, the solutions to manage growth of the plant in common lands and water bodies require more than removing plants at one go.

Judicial environmentalism

The law

The Indian Forest Act 1927 identifies prosopis and its various products, such as fuelwood, timber, charcoal and flowers, as forest produce but does not bar anyone from planting the species and enjoying the products. In most states, including Tamil Nadu and Andhra Pradesh, cutting and selling prosopis from private property does not require prior permission of the landowner. Such control applies only to tank beds, bunds and other common lands listed for maintenance, management and control by selected government departments. These include the Public Works, Rural Development and Local Administration, Revenue and Forest Departments. Many disputes between wood contractors and regulators (such as the Revenue and Forest Departments) have reached the courts, especially with respect to establishing procedures for auctioning and for realizing revenues from common lands such as tank beds and river channels (e.g., *K. Thandeeswaran v The Principal Chief Conservator*, 2014). The extant law, including both statutes and court decisions, considers prosopis as a valuable plant for both public and private interests. Only a few states, such as Gujarat, have strictly regulated cutting and transportation of prosopis trees, even from private lands (MoEFCC (Ministry of Environment, Forest and Climate Change), 2012). Parties seeking to gain small concessions for transporting prosopis wood and charcoal by farmers are resisted rigorously by the department in the courts (*Bhimji Jethabhai Vador v State of Gujarat*, 2014).

Apart from state government departments, many organized village communities in one or another way claim collective ownership over prosopis plantations on common lands such as tank beds and bunds. For example, in the districts of Madurai, Ramanathapuram and Tirunelveli in Tamil Nadu, many communities collect the income derived from prosopis contractors.

The strength of community-based institutions in a given village is measured in terms of how big their tank is and the value of standing prosopis trees on common lands at any given point in time. In many villages, people collectively desist from removal of prosopis trees ordered by the court or other government authorities, in order to protect future stream of revenue for their village functions and temple celebrations.

In short, the law does not define prosopis as harmful through any extant quarantine, biodiversity or other legislation in India. Rather, the law considers prosopis a useful and economically valuable plant for individuals and the state. Nor is there any provision in the law that prohibits cultivation of the trees or enjoyment of the wood, shade and other benefits from the plant by anyone, including private landowners and the landless who depend on common lands.

Intervention by constitutional courts

Through public interest litigation, activists often approach India's constitutional courts – the Supreme Court and the state high courts – to seek remedies in resolving environmental concerns. In the recent past, petitioners have brought cases seeking writs to order the government to address and remedy a wide array of environmental damages. In the last 30 years, issues that have been litigated include the decay of the natural environment in hills, pollution and encroachment in rivers and water bodies, rampant water exploitation and removal of exotic plants from reserved forests. Two such prominent cases were fought in the Madras High Court and dealt with removal and control of exotic plants, such as wattle and eucalyptus, from the Nilgiris (*K. Ussainar v The State of Tamil Nadu*, 2014) and Kodai (*Saravanan v Secretary Forest Department, Tamil Nadu*, 2015) hills in Tamil Nadu.

With respect to prosopis, Vaiyapuri Gopalsamy (known as Vaiko), political leader of a prominent regional party, and a few activists approached the Madras High Court between 2015 and 2017, seeking writs ordering the government to remove the plant from riverbeds, tanks, channels and all types of private lands. In general, these petitions claimed that the plant had entered water bodies and channels apart from fallow private lands. The environmental concerns about prosopis – for example, that it depletes groundwater and affects other native plants – were cited in these petitions. None of these claims were supported with scholarly studies or reports, and the petitioners did not present various benefits of prosopis. Thus, the court did not hear a balanced argument about the impacts and benefits of the plant.

The court agreed with many of the arguments presented by the petitioners and ordered the government to assess and remove the plant from riverbeds and later enlarged the scope to all water bodies such as tanks, road margins and common lands in 13 districts of Tamil Nadu that fell under its jurisdiction. Municipal corporations were directed to remove the plant from their spaces and facilities, and households were required to remove it from their surroundings (*The Hindu*, 2016). Massive campaigns were launched by government departments in towns and villages, citing this interim order. The court drew up a scheme to monitor implementation of its orders by advocate commissioners and judicial officers drawn from the lower judiciary (*Vaiko v Chief Secretary and 738 Others*, 2017). The order was addressed to some 739 respondents, including all departments that have some responsibility for owning or managing land within Tamil Nadu, the chief secretary of the state, two principal secretaries, 32 district collectors, 219 tehsildars (revenue department officials), 383 block development officers, defence estate officers, naval commanding officers, port officials, airport officials of all airports and highway research stations. The judges were very displeased and unsatisfied with the "insouciant" behaviour of district officers regarding the initial ruling to remove prosopis and expressed that in their order. Certainly, this order is remarkable by any measure. Furthermore, the district judges were given targets to undertake visits and coordinate with the district collectors to oversee the removal process (Imranulla, 2017a, 2017b, 2017c, 2017d, 2018). Some of the judicial officers even went so far as to organize special camps to undertake the removal of prosopis using local volunteers. This action by the court raises many questions about the role of the courts in the Indian judicial and legal systems.

From December 2016 to March 2017, judges from the Madurai Bench of the Madras High Court made field visits to check how their orders were being implemented (Imranulla, 2017a, 2017b, 2017c, 2017d, 2018). District collectors, senior engineers and many officers from various government departments accompanied the judges in a convoy of vehicles to explain their activities to the judges. Similar visits by district judges, sub-judges and groups of advocate commissioners were reported across southern districts in Tamil Nadu. Inspections of this nature made by judges are rare and unheard of in the judicial history of Tamil Nadu. Over 200 advocate commissioners were appointed by the high court to monitor implementation of the court order. The administration disseminated information about the court order through advertisements in print media and on radio programmes. The City Corporation of Madurai and many other municipalities in the state displayed hoardings (billboards) to instruct citizens about the court orders and encourage them to remove prosopis from their property or else face legal penalties. Judges even suggested that district collectors use the Tamil Nadu Revenue Recovery Act 1890 to collect the cost of removal from landowners who fail to comply with the order.

In one such encounter, a vernacular newspaper reported that

> [j]udges questioned the officials why the trees are not removed from tank beds. The officers [from the line departments] replied "It is in the government lands and hence we have called for tenders to dispose of them … In several villages, the villagers are grooming and safeguarding them to raise revenue for themselves. Therefore, they object to removal

and hence the delays ..." The officers requested the judges to deploy machines to remove its thick growth and wanted 90 days more to remove them in another 40,000 ha of lands ... Judges responded prosopis should be removed without any interruptions and said the court will give its support at all times ... Thereafter the judges went to Melakonnaikulam village ... and spoke to villagers and told them "These trees spoilt the groundwater, environment, and agriculture on a large scale. Therefore, they must be removed with its roots. Such removal will help future generations. Removing prosopis must be converted into a Peoples Movement.

(*Dinamani*, 2017)

The same news report said that local villagers objected to contractors being sent by local government authorities to remove prosopis from their village tanks. The villagers claimed that the revenue that would have otherwise come to the village from harvesting the plant would be taken away by such contractors.

In another interim order, the judges directed the state government "to enact a special legislation within the next couple of months, with prohibitory and penal clauses for eradicating prosopis that is considered harmful to the environment – from public and private lands alike" (*Vaiko v Chief Secretary and 738 Others*, 2017). Judges of the Madurai Bench of the Madras High Court were so enthused that they created a fund to collect money to promote the cause of prosopis eradication. Many of the district judges and a few high court judges even made removal of prosopis a condition for issuing bail in criminal cases (Jesudasan, 2017). In one such case, the petitioner seeking bail filed a separate court case challenging the bail condition requiring removal of 100 prosopis trees in 20 days (*Gnanam v Inspector of Police (L&O) Chennai*, 2017). After hearing the arguments in this case, the judge observed, "The present spree or competition among [some] the Judges in our State to impose such bail conditions [of removing prosopis] signals not march of law but an onslaught on human rights, human elements, and human sentiments" (*Gnanam v Inspector of Police (L&O) Chennai*, 2017, para 42).

Impact of the judgement and judicial activism

Except for the Revenue Department, none of the respondents that were implicated in this case were parties to the original lawsuit. Their views were never heard by the court, and yet sweeping orders were passed that affected them and had profound resource and legal consequences. Furthermore, no scientist or research agency participated in the proceedings to offer scientific opinions. At the very least, the judges should have reminded themselves to clearly identify the legal foundation for their powers in handling an issue affecting so many people, offices and establishments. However, they did not care to ensure that this basic requirement was met before they arrived at their decision, in which they also ruled that they would inspect the removal works themselves, along with the district judges and advocate commissioners.

In the meantime, following the judges' activism, the price of wood crashed from around the end of January 2017, dropping from Rs. 2,400 to Rs. 1,600 per tonne. Although high-quality charcoal can be produced from root stubs, the price of stubs came down from Rs. 2,300 to Rs. 1,000, as there were not many takers for producing charcoal. This was because the price of charcoal fell from Rs. 13,000 to Rs. 6,000 per tonne. Traders claimed there was a glut in the market because of overproduction, based on court orders. The low prices and glut in the market remained for more than a year until the government started ignoring the court order.

Removal of prosopis from the Indian Institute of Technology's Madras campus led to disturbances to the blackbucks that are endemic to the area, and the Forest Department thus instructed the institute not to remove trees if doing so would disturb blackbuck habitat. Hence, the removal of prosopis trees was stopped (Sujatha, 2017). The First Bench of the Madras High Court ordered the operations to be stopped until a scientific basis for removing prosopis was clarified (TNN, 2017). Thereafter, the momentum of the wholesale removal of the plant came to a halt in most parts of the state, because of a change in public opinion (Thirumurthy, 2017) and because several civil and property disputes had arisen between government departments and landowners, contractors and other prosopis users (e.g., *Mangathayammal v District Collector and 6 Others*, 2017). However, the order of the Madurai Bench of the Madras High Court to remove the plant still remains on the books, and occasionally judges refer in open court to implementing the order. In 2018, at one such hearing, the judges asked why their prosopis orders had not yet been carried out and wanted to revive the monitoring process established in their order of February 2017 (Imranulla, 2018). As recently as 2020, the Madras Bench of the High Court once again brought forward the prosopis issue and appointed a scientific commission to study the "ill effects" of the plant (*The Hindu*, 2020). It is time to seek answers from the courts about the fairness and usefulness of such judicial activism when it is not fully based on solid scientific evidence and sound legal reasoning.

The above discussion shows that the decisions of the court and actions of the judges are worrying and unwarranted, as they touch thousands of people. First and foremost, the decision lacks a scientific basis and legal rationale for removing the plant in order to solve the environmental problems that it supposedly causes – groundwater depletion and environmental and agricultural damages. The layperson in rural areas does not share the judges' concerns about prosopis, and there is a lack of scientific evidence for the sweeping claims made about the harm caused by the plant. Second, what is the "civil wrong" the judges found with those farmers who neither planted prosopis nor wanted it removed from their fields that would justify forcing them to bear the cost of removal, especially when so many farmers live in poverty and make their living on prosopis products? Third, how can judges use the Revenue Recovery Act to order authorities to collect the cost of removing prosopis from farmers' fields? That law is normally used to make recoveries from landowners who are in arrears in taxes or other government dues owing on land revenue. Using the act for this other purpose subverts basic property rights over the land and livelihood opportunities. Farmers who neither planted nor wanted to have prosopis in their fields are now financially responsible for the plant's removal and face serious consequences if they do not pay. The courts have no sound legal or equitable basis to fix this responsibility upon farmers, many of whom are already in distress. Fourth, what is the legal rationale or the constitutional validity of this judicial environmentalism? The constitutional scheme of the country does not allow for the type of active implementation role played by the judiciary in this case. Such a function is usually left to various enforcement arms of the executive. Finally, why did the courts show such activism after hearing only one perspective, that of the petitioners, before making a decision, but have not taken such drastic measures in similar situations involving equally serious environmental issues, such as sand mining, quarrying and polluting? The use of public interest litigation and subsequent activism of courts in the case of prosopis is highly unusual and worrisome, as the cost of "cleaning or reclaiming the environment" is disproportionately imposed on the poor. Both the activists and various levels of court involved in prosopis cases in Tamil Nadu have defeated the fundamental purpose of judicial activism by failing to account for all four components of environmental justice for the poor.

References

Baskin, Yvonne. (2002) *A Plague of Rats and Rubbervines: The Growing Threat of Species Invasions.* A SCOPE-GISP Project. Washington, DC: Island Press.

Bhimji Jethabhai Vador v State of Gujarat. (2014) Gujarat High Court WP No. 326 of 2013, decided on 17 January. Available at: https://indiankanoon.org/doc/59538809/.

CABI. (2019) *Prosopis juliflora (Mesquite), Invasive Species Compendium.* Wallingford: CAB International. Available at: www.cabi.org/isc/datasheet/43942 (Accessed 21 December 2020).

Dinamani. (2017) "Hasten the work to remove prosopis trees: Judges instructed Sivagangai District collector" (in Tamil), *Dinamani,* 23 February.

Edmond, Deepu Sebastian. (2017) "Even as Tamil Nadu battles the Karuvelam, Rajasthan tries to work with it," *The Hindu,* 15 November. Available at: https://www.thehindu.com/news/national/even-as-tamil-nadu-battles-the-karuvelam-rajasthan-tries-to-work-with-it/article20460318.ece.

Gnanam v Inspector of Police (L&O) Chennai. (2017) High Court of Madras Criminal Original Petition No. 4530 of 2017, CDJ 2017 MHC 1144. Available at: https://www.legitquest.com/case/gnanam-v-state-rep-by-the-inspector-of-police-l-o-chennai/A404F.

Government of Tamil Nadu. (2014) *Season and Crop Report (2013–14).* Chennai: Department of Economics and Statistics.

The Hindu. (2016) "Corporation wields the stick against 'seemai karuvelam,'" *The Hindu,* 24 December. Available at: https://www.thehindu.com/news/cities/Tiruchirapalli/Corporation-wields-the-stick-against-%E2%80%98seemai-karuvelam%E2%80%99/article16936925.ece.

The Hindu. (2020) "Seemai Karuvelam: HC ropes in NEERI," *The Hindu,* 18 February. Available at: https://www.thehindu.com/news/cities/chennai/seemai-karuvelam-hc-ropes-in-neeri/article30846703.ece.

Imranulla, Mohamed S. (2017a) "Principal District Judges asked to inspect removal of 'seemai karuvelam,'" *The Hindu,* 10 January. Available at: http://www.thehindu.com/news/national/tamil-nadu/Principal-District-Judges-asked-to-inspect-removal-of-%E2%80%98seemai-karuvelam%E2%80%99/article17019032.ece.

Imranulla, Mohamed S. (2017b) "HC asks Principal District Judges to inspect removal of 'seemai karuvelam,'" *The Hindu,* 12 January. Available at: http://www.thehindu.com/todays-paper/tp-national/tp-tamilnadu/HC-asks-Principal-District-Judges-to-inspect-removal-of-%E2%80%98seemai-karuvelam%E2%80%99/article17025912.ece.

Imranulla, Mohamed S. (2017c) "Judge expresses dissatisfaction over progress in removal of seemai karuvelam," *The Hindu,* 31 January. Available at: http://www.thehindu.com/news/cities/Madurai/Judge-expresses-dissatisfaction-over-progress-in-removal-of-seemai-karuvelam/article17118705.ece.

Imranulla, Mohamed S. (2017d) "HC judges make surprise inspection of seemai karuvelam removal," *The Hindu,* 20 February. Available at: http://www.thehindu.com/news/national/tamil-nadu/HC-judges-inspect-seemai-karuvelam-in-Madurai/article17335888.ece.

Imranulla, Mohamed S. (2018) "HC revives crusade against seemai karuvelam trees," *The Hindu,* 26 December. Available at: https://www.thehindu.com/news/national/tamil-nadu/hc-revives-crusade-against-seemai-karuvelam-trees/article25829012.ece.

Jesudasan, Dennis S. (2017) "Novel bail conditions raise questions of propriety," *The Hindu,* 20 March. Available at: http://www.thehindu.com/news/national/tamil-nadu/novel-bail-conditions-raise-questions-of-propriety/article17533132.ece.

K. Thandeeswaran v The Principal Chief Conservator. (2014) Madurai Bench of Madras High Court WP (MD) No. 16878 of 2014, decided on 19 November 2014, CDJ 2015 MHC 596. Available at: https://indiankanoon.org/doc/133296090/.

K. Ussainar v The State of Tamil Nadu. (2014) Madurai Bench of Madras High Court WP No. 16857 of 1991, decided on 17 December 2014, CDJ 2015 MHC 065. Available at https://indiankanoon.org/doc/16998062/.

Mangathayammal v District Collector and 6 Others. (2017) Madurai Bench of Madras High Court WP (MD) No. 4499 of 2017, decided on 17 March 2017, CDJ 2017 MHC 1242. Available at: https://indiankanoon.org/doc/198549275/.

MoEFCC (Ministry of Environment, Forest and Climate Change). (2012) *The Regulatory Regime Regarding Felling and Transit Regulations for Tree Species Grown on Non Forests/Private [sic] Lands.* New Delhi: Ministry of Environment, Forest and Climate Change, Government of India. Available at: http://www.moef.nic.in/assets/Rep_Committee_Trees_on_Private_land-27122012.

Malhotra, S. and Misra, K. (1983) "A novel tannin from *Prosopis juliflora* roots," *Current Science,* 52(12), 583–585.

Mosse, David. (2003) *The Rule of Water: Statecraft, Ecology and Collective Action in South India.* New Delhi and Oxford: Oxford University Press.

Narayanan, R.S. (2014) "No problem with the trees [Marangalidam Thavarillai]" (in Tamil), *Dinamani,* 31 March 2014.

Palanisami, K. and Balasubramanian, R. (1995) "Resource mobilisation and tanks performance under panchayat and PWD management in Tamil Nadu" in *National Workshop on Traditional Water Management for Tanks and Ponds,* Madras vol. Madras: CWR, Anna University.

Pasiecnik, N.M., Felker, P., Harris, P.J.C., Harsh, L.N., Cruz, G., Tewari, J.C., Cadoret, K. and Maldonado, L.J. (2001) *The Prosopis Juliflora: Prosopis Pallida Complex: A Monograph.* Coventry: HDRA. Available at: https://www.gardenorganic.org.uk/sites/www.gardenorganic.org.uk/files/resources/international/ProsopisMonographComplete.pdf.

Reddy, Konda C.V. (1978) "*Prosopis juliflora*, the precocious child of the plant world," *Indian Forester*, 104(1), 14–18.

Sakthivadivel, R. (2016) "*Prosopis juliflora* in the irrigation tanks of Tamil Nadu," *Water Policy Research Highlight*, 7, 1–6. Anand, Gujarat: IWMI Tata Water Policy Program. Available at: http://www.iwmi.cgiar.org/iwmi-tata/PDFs/iwmi-tata_water_policy_research_highlight-issue_07_2016.pdf.

Saravanan v Secretary Forest Department, Tamil Nadu. (2015) Madurai Bench of Madras High Court WP (MD) No. 3633 of 2014, decided on 10 August 2015, CDJ 2015 MHC 7796.

Seenivasan, R. (2014) *Law, Technology and Water Conflicts in Developing Societies: Case Study of Tank Systems of Tamil Nadu.* PhD thesis, University of Westminster, London. Available at: http://westminsterresearch.wmin.ac.uk/15583/.

Seenivasan, R. and Kumar, Anand. (2004) *Vision for Village Tanks of India*, vol. 1. Madurai: DHAN Foundation.

Singh, Arvind. (2007) "Revegetation of coal-mine spoils using *Prosopis juliflora* in Singrauli coalfield is a harmful practice from an ecological viewpoint," *Current Science*, 93(9), 1204.

Sujatha, R. (2017) "'Stop removal of seemai karuvelam from IIT-M,'" *The Hindu*, 12 April. Available at: https://www.thehindu.com/news/cities/chennai/stop-removal-of-seemai-karuvelam-from-iit-m/article17935771.ece.

Tewari, J.C., Mathur, B.K., Tewari, P., Singh, Y., Singh, M., Ram, M. and Sharma, A. (2013) "*Prosopis juliflora*: A miracle species of hot arid and semi-arid regions of India," *Popular Kheti*, 1(2), 53–60.

Thirumurthy, Priyanka. (2017) "Is the Karuvelam tree an ecological threat? Experts divided as Madras HC stays its eradication," *The News Minute*, 5 January. Available at: https://www.thenewsminute.com/article/karuvelam-tree-ecological-threat-experts-divided-madras-hc-stays-its-eradication-61301.

TNN. (2017) "No scientific basis – High Court stays battle against juliflora trees," *Times of India*, 29 April.

Vaidyanathan, A. (2001) *Tanks of South India*. New Delhi: Centre for Science and Environment.

Vaidyanathan, A. (2006) *India's Water Resources: Contemporary Issues on Irrigation*. New Delhi: Oxford University Press.

Vaiko v Chief Secretary and 738 others. (2017) Madras High Court WP (MD) No. 16485 of 2015, Interim Order dated 27 February 2017. Available at: https://www.casemine.com/judgement/in/5ac5e4ff4a93261ae6b41625.

Voelcker, John Augustus. (1893) *Report on the Improvement of Indian Agriculture [With Map]*. London: Eyre & Spottiswoode.

Walter, Kurt J. and Armstrong, Karen V. (2014) "Benefits, threats and potential of *Prosopis* in South India," *Forests, Trees and Livelihoods*, 23(4), 232–247. Available at: https://doi.org/10.1080/14728028.2014.919880.

Part IV

Conclusion

15 Lessons for policy and institutional reform

Alan P. Diduck, Kirit Patel and Aruna Kumar Malik

India's institutional and policy framework, at both central and state levels, includes important advances and innovations which, although flawed, have potential to promote environmental justice for marginalized communities. Examples include mechanisms that support public interest litigation (PIL), specialized environmental courts and tribunals, requirements for environmental impact assessment (EIA) and an array of environmental regulatory agencies. In no small measure, the effectiveness of these institutions and policy developments has been held back by neoliberal economic priorities – such as the ease of doing business approach adopted by the central government and the privatization and deregulation or reregulation of small industries. Within this socio-economic context, the book has applied concepts from the broader environmental justice literature – recognitional, procedural, distributive and restorative justice – to inquire into and critique selected aspects of India's institutional and policy framework that have implications for environmental protection, and social justice. This chapter provides an overview and synthesis of selected findings and recommendations presented in each of the book's chapters and offers concluding thoughts for the way forward.

Key findings and recommendations

Key results from each of the chapters shed light on how socio-economic contextual factors or elements of the institutional and policy framework have constrained one or more of the recognitional, procedural, distributive and restorative dimensions of environmental justice. The results also reveal promising avenues for overcoming these constraints, including in some cases specific recommendations for institutional and policy reforms. Important examples of the impacts and implications of socio-economic contextual factors are found in Chapters 2 and 3. Here we see how the totalizing effects of neoliberal policies can compound lack of political attention to recognitional justice, justify procedural injustices and help whitewash blatant inequities in the distribution of environmental benefits and burdens. The ultimate impacts described in those chapters include dispossession of common or community property and disregard for statutorily protected rights of prior consent. At the same time, these chapters reveal the potential that local resistance holds for defending rural environments and livelihoods; promising models of

DOI: 10.4324/9781003141228-15

collective governance of small-scale hydro; and intriguing prospects for local resistance coupled with jurisdictional leapfrogging to the international level.

Chapters 4–6 highlight the impacts and implications of crucial components of the institutional and policy framework. In the context of water justice, Chapter 4 shows how common law principles governing the water–land nexus block access to groundwater for those who do not own real property. It also reiterates the promise and potential of remedies based on Article 21 of the Constitution of India and landmark decisions, such as *Subhash Kumar v State of Bihar* (1991), that have extended Article 21 to include environmental amenities. What is still needed is statutory recognition of the right to water, recognitional justice to ensure that the voices of all rights holders are heard and improvements to procedural justice, including better rights to information, public participation and access to the courts.

Recognitional and procedural justice are also at the heart of Chapter 5, which focuses on the interests of two highly marginalized communities – tribal peoples and fisherfolk. The chapter canvasses a selection of decisions made by the National Green Tribunal (NGT), a significant element of India's institutional framework governing environmental matters and one meant to further procedural justice in environmental disputes. Additionally, the chapter reveals how heeding recognitional justice and improving access to specialized judicial forums can improve equity in the distribution of environmental burdens and benefits and advance aspects of restorative justice, such as rehabilitation and compensation. Analysis of a range of NGT decisions involving tribal peoples and fisherfolk leads to the conclusion that overall, the decisions reflect a pragmatic and balanced approach when considering economic and environmental interests.

Chapter 6 focuses on a crucial element of India's policy framework, the Environment Impact Assessment (EIA) Notification 2006 and in particular the expert appraisal phase. A review of pertinent provisions of the notification as well as judicial and NGT decisions raises serious concerns with the appraisal process as well as with judicial oversight of the process. One issue is that EIA documents have been inadequately vetted during appraisal; another is the judicial leniency given to project proponents and regulatory authorities who are in violation of their statutory duties. Both of these problems obviously pose major impediments to achieving distributive and restorative justice. The chapter recommends several solutions, including increasing the financial and human resources devoted to appraisal and encouraging the judiciary to take a stricter approach with delinquent project proponents and regulatory authorities, especially with respect to material or incurable defects in appraisal. To these solutions, we would add a recommendation for increased opportunities and capacity for public vetting of EIA documents. This suggestion is described in more detail below, in relation to Chapters 8–10 and 12.

Supplementing the political economy analysis found in Chapters 2 and 3, Chapter 7 provides a close examination of five small-scale hydro projects in the Kullu District of Himachal Pradesh and reveals problems with the environmental approval and clearance processes for such projects. A key finding is the inequitable distribution of the costs and benefits of small hydro: benefits largely accrue to private sector investors and the state and regional economies in general, while adverse impacts are mostly felt at the local level. These impacts include effects on aquaculture, traditional milling livelihoods, irrigation potential and river corridor ecology and hydrology.

Part of the solution is that small hydro proponents and regulators need to better recognize the interests, knowledge and acumen of local mountain people and establish meaningful ways for them to participate in planning and decision-making. Another part of the solution is to conduct an inclusive strategic impact assessment of small hydro policy that would act as an overall framework for decisions about individual projects. Additionally, periodic inclusive regional or catchment-based cumulative effects assessments should be mandatory. These changes would allow for long-term planning and thoughtful consideration of future impacts of a proposed project in combination with impacts already present.

Chapters 8–10 consider different types of industrial development in Gujarat. Chapter 8 examines a limestone mine in Bhavnagar District, Chapter 9 centres on a cement factory – also in Bhavnagar, while Chapter 10 focuses on a coal-fired power plant in Kutch District. The chapters share some common results and lessons for improving environmental justice: each supplements and confirms the lessons of Chapters 2 and 3 regarding the power of neoliberal economic policies to restrict efforts to achieve environmental justice. This is also true of Chapter 12, which analyses the environmental clearance granted for a major port expansion in Karnataka. Although none of these chapters offer an explicit political economy analysis, it is quite clear that in each case, ease of doing business or similar policy priorities were underlying economic drivers. Such policies have long been in effect in Gujarat. The power plant in Kutch (Chapter 10) is located in a Special Economic Zone established by the Adani Group, a multinational conglomerate, while the Karnataka port expansion project (Chapter 12) is part of the Sagarmala programme launched by the federal Ministry of Shipping to modernize port infrastructure and enhance national capacity to import and export goods.

Another common feature of the case studies in these four chapters (Chapters 8–10 and 12) is that each reveals serious flaws in the impact assessment process established by the EIA Notification, most notably inadequate opportunities for public engagement and faulty impact assessment methods and reporting. Coupled with significant potentially adverse local impacts, these process flaws have led to conflict, litigation and widespread social opposition, including agitation and other extra-legal measures. Chapter 10 shows how local resistance can be coupled with jurisdictional leapfrogging, similar to what is detailed in Chapter 4. More particularly, with supports from EarthRights International, local groups objecting to the power plant in Kutch were able to advance their cause on the international stage and pursue remedies in the American courts. These efforts by the fishing community and Machimar Adhikar Sangharsh Sangathan activists indicate their frustration with Indian regulatory agencies and judicial institutions. Though one cannot predict an outcome of litigation in the US courts, if such legal action was to be successful, it would create a new and promising avenue for holding international financing institutions accountable. It would establish a new mechanism to force financing institutions to comply fully with their institutional policies for assessing environmental risks and consequences.

An important lesson flowing from these four chapters is the need for increased opportunities and capacity for public vetting of impact assessment methods and reports. The public engagement procedures mandated by the EIA Notification need to be made more inclusive and transparent. Locally affected communities need early and ongoing opportunities to be involved in project assessment, rather than just being invited to attend a one-off public hearing and to review documents on a government website. Such policy enhancements, complemented by increased attention to recognitional matters such as local interests, needs and aspirations, would significantly improve the prospects for procedural justice. Furthermore, public hearings need to be made more deliberative and formal, with adequate resources provided to carefully chosen interveners, so they can participate fully and muster sound evidence and credible arguments that supplement government appraisals of positions proffered by project proponents. Participant assistance of this type is a basic ingredient of meaningful participation (Sinclair and Diduck, 2021) and has proven to be effective in jurisdictions where it has been implemented (e.g., Impact Assessment Agency of Canada, 2020). A reform like this would help protect the interests of locally affected communities and advance the broad public interest in environmental protection and equitable economic development, thus helping to prevent distributive inequities.

Chapter 9 also reveals the potential for gender-based analysis, increasingly used in impact assessments around the world (e.g., Walker and Reed, 2021), to improve the capacity of EIA to advance recognitional justice and improve distributive equity. This chapter reiterates how the neoliberal development narrative not only disregards the impacts of environmental degradation on the livelihoods of rural women but also ignores the contributions women make in seeking justice. Thus, gendered analysis is not only important in EIA processes led by project proponents but also in various protests and resistance organized by environmental justice movements. Without recognizing the many roles women play, it is difficult to ensure that they receive justice.

Chapter 12 highlights how the failure to consider recognitional and distributive justice prevents the advancement of restorative justice, and recommends heightened attention to the restorative dimension when regulatory authorities establish conditions in environmental clearances. In Chapters 8–10 and 12, we also saw that PIL had mixed results, but in the mining and cement cases, the court challenges at least had the effect of stalling project implementation. Stalling implementation, especially at advanced stages, is obviously not ideal for the environment or society at large; however, in these cases, the delays permitted local communities to forge ahead with the social movements in opposition to the projects.

Similarly, Chapter 13 reports mixed results from PIL in the matter of the Sterlite Copper smelter in Thoothukudi. Although the smelter was ultimately closed, this can be attributed to its extensive history of regulatory non-compliance, widespread public protests and the police action in 2018 that left several protesters dead, rather than a single court decision. Other than the tragic loss of human life, perhaps the most disconcerting parts of the Sterlite case might be the post-facto environmental clearances that were granted for some production operations and the frequent lack of enforcement on the part of the regulatory authorities. Far too often, Sterlite was able to act with impunity when it was in violation of a regulatory or environmental clearance requirement. In some instances, this problem was compounded by the NGT's refusal to use its special power to investigate matters on site or to compel the Tamil Nadu Pollution Control Board to enter into a full investigation. An important recommendation from Chapter 13 is for state governments to bolster their regulatory tools and enhance the human and financial resources available to pollution control boards and other regulatory agencies. The enforcement and compliance gap must be closed to prevent distributive and restorative injustices and help offset vexatious proceedings by corporate interests.

To some degree, Chapter 11 presents a success story, reporting on PIL that resulted in the NGT's overturning of sand mining permits because of procedural irregularities in the permit application process. Procedural justice was, for the most part, effectuated in this case and potential long-term distributive injustices might be avoided. However, similar to the case in Chapter 13, a key finding here was that enforcement of the NGT decision was lacking, and unless rectified, the tentative procedural and distributive victories in this case will be undermined. Additionally, the challenges faced by local residents in trying to access the NGT system brought to light important ways to improve PIL procedures and thereby promote procedural justice. Public financial support to help retain legal counsel and technical experts should be available to petitioners who otherwise would not have the means to bring forth claims that advance legitimate public interests. The opportunity to act in the broad public interest or protect local environments and livelihoods should not be reserved for those with deep pockets or easy access to pro bono legal assistance.

Another side of PIL is presented in Chapter 14. Here we see what is arguably a drastic case of judicial overreach in a series of decisions of the Madras High Court ordering removal of the invasive plant *Prosopis juliflora*. The decisions included unusually intrusive directives regarding enforcement and compliance and reflected a lack of the integrative and balanced mindset required to fully consider environmental problems with widespread social and economic implications. A single-minded focus on eradication of the species resulted in a failure to consider what eradication – and the methods of eradication – would mean for rural livelihoods, common pool resources and property rights. Environmental protection and improvements were prioritized at the expense of distributive and restorative justice. What was obviously needed here was greater attention to recognitional issues during the PIL proceedings and a wider array of perspectives heard by the court. The case proceeded without direct involvement of the wider community, including people who would be directly affected. The judiciary needs to ensure participation of affected parties instead of adjudicating with only limited input from litigants or relying on their own assumptions about environmental degradation, rural livelihoods and poverty.

Concluding thoughts

The key findings summarized above include important examples of policy and institutional innovations adopted in India over the years, including PIL, EIA and the NGT, that have helped protect and advance environmental justice for marginalized communities. At the same time, the results and associated policy suggestions show quite clearly that further reforms and innovations are not only possible but necessary. The cause of environmental justice has been held back by imperatives associated with neoliberal policies adopted in the name of rapid economic growth. More balance is obviously needed in national and state planning between growth-oriented policies on the one hand and sustainability focused policies that have due regard for all aspects of environmental justice on the other. Additionally, important gaps exist in the policy and institutional regime that impede progress on environmental justice, although promising and feasible suggestions for reform are identified. For example, incorporating gender-based analysis into EIA would advance recognitional justice. Developing formalized mechanisms for providing legal and other technical resources to EIA public intervenors and PIL petitioners would enhance prospects for procedural justice. Legislating a right to water and strengthening EIA appraisals would help promote distributive equity. Crafting meaningful environmental condition clearances, bolstering regulatory enforcement capacities and instituting well-resourced systems of community-based monitoring would help prevent distributive inequities as well as advance restorative justice.

However, despite their promise and practicability, the pursuit of these reforms and others like them would undoubtedly face serious challenges. Most require legislative change, some may need judicial approval, and to be successful, all require considerable political support. In the current socio-economic climate in India, mustering that support would be a formidable task. For instance, the central government's Draft EIA Notification 2020 (MoEFCC (Ministry of Environment, Forest, and Climate Change), 2020) represents a retreat from many of the suggested reforms highlighted above. If adopted, it would provide a blanket exemption from the public hearing requirement for projects that fall into a newly created B2 category. It would also exempt expansions of less than 50 per cent of current capacity in existing projects. Furthermore, the Draft EIA Notification would reduce the notice period for public hearings, reduce periodic compliance requirements and institutionalize post-facto clearances (Gupta, Nayak, Tanvani and Viswanathan, 2020). The spirit of the proposed amendment is clearly to speed up the environmental clearance process, yet another example of re-regulation in support of the ease of doing business model. If approved, it will be a huge step back for environmental justice. Moreover, the efforts of the neoliberal government are not limited to undermining the EIA regime. The NITI Aayog, the government's apex think tank and economic planning branch, has recently commissioned a study to examine the economic consequences of judicial decisions that have hindered large projects on environmental grounds (Koshy, 2021). Though the NITI Aayog has limited capacity to influence the constitutional courts, it can certainly impact the functioning of tribunals and special courts such as the NGT.

Despite these challenges, what gives hope are the examples cited throughout the book of collective governance, social movements and non–governmental organization and action. Such oppositional efforts are, of course, an enduring characteristic of India's modern political history. The successes of groups such as the Sai Engineering Foundation and the Churah Floriculture Cooperative Society reiterate the potential of alternative forms of business organization, especially in the context of small hydro development. Movements such as the Mahuva Andolan and Niyamgiri Suraksha Samiti demonstrate the benefits of bypassing formal policy and institutional measures when fighting for environmental justice. Faced with roadblocks in the formal system, these groups have achieved success by resorting to political pressure and civil disobedience, including in some cases agitation and other extra-legal measures. Their experiences also, of course, reveal the perils of adopting such tactics. Movement leaders have incurred serious financial, opportunity and other costs, have been put in legal jeopardy and in some cases physically assaulted. These perils underscore the need for ongoing improvements in the policy and institutional framework, at both central and state levels. Finally, groups such as the trade association Machimar Adhikar Sangharsh Sangathan and initiatives such as the Environmental Justice Program supported by the Centre for Policy Research and Namati substantiate the power of community organizing and local capacity development. Their work to protect rural livelihoods and empower laypeople to pursue legal remedies in protection of local environments is inspirational. Continued community-focused efforts like these, coupled with policy and institutional reforms such as the ones proposed above and more sustainability focused economic planning by government, would go a long way to promoting and protecting environmental justice in India.

References

Gupta, D., Nayak, S., Tanvani, K. and Viswanathan, V. (2020) *The Draft EIA Notification 2020: Reduced Regulations and Increased Exemptions (Part 1)*. New Delhi: Centre for Policy Research-Namati Environmental Justice Program. Available at: https://

cprindia.org/sites/default/files/Reduced%20Regulations%20and%20Increased%20 Exemptions_Part%20I_30.07.pdf (Accessed: 19 February 2021).

Impact Assessment Agency of Canada (2020) *Funding Programs*. Impact Assessment Agency of Canada. Available at: https://www.canada.ca/en/impact-assessment-agency/services/ public-participation/participant-funding-application-environmental-assessment.html (Accessed: 20 February 2021).

Koshy, J. (2021) "NITI Aayog study to track economic impact of green verdicts," *The Hindu*, 7 February 2010. Available at: https://www.thehindu.com/news/national/ niti-aayog-study-to-track-economic-impact-of-green-judgements/article33770515. ece (Accessed: 19 February 2021).

MoEFCC (Ministry of Environment, Forest, and Climate Change) (2020) *Environment Impact Assessment Notification 2020. Ministry of Environment, Forest, and Climate Change.* Available at: https://environmentclearance.nic.in/writereaddata/Draft_EIA_2020. pdf (Accessed: 20 February 2021).

Sinclair, A.J. and Diduck, A.P. (2021) "Considering the potential for meaningful public participation under Canada's new Impact Assessment Act" in Doelle, M. and Sinclair, A.J. (eds.), *The New Canadian Impact Assessment Act (IAA)*. Toronto: Irwin Law.

Subhash Kumar v State of Bihar (1991) AIR SC 420.

Walker, H. and Reed, M.G. (2021) "Assessing the intersections of sex, gender, and other identity factors in the new Canadian Impact Assessment Act" in Doelle, M. and Sinclair, A.J. (eds.), *The New Canadian Impact Assessment Act (IAA)*. Toronto: Irwin Law.

Index

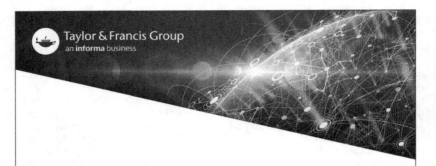

Printed in the United States
by Baker & Taylor Publisher Services

Printed in the United States
by Baker & Taylor Publisher Services